Single-Cell Analysis

Methods and Protocols

Edited by

Sara Lindström

Department of Cell and Molecular Biology, Science For Life Laboratory, Karolinska Institute, Stockholm, Sweden

Helene Andersson-Svahn

Division of Nanobiotechnology, AlbaNova University Centre, Royal Institute of Technology, Stockholm, Sweden

Humana Press

Editors
Sara Lindström
Department of Cell and Molecular Biology
Science For Life Laboratory
Karolinska Institute
Stockholm, Sweden
sarali@kth.se

Helene Andersson-Svahn
Division of Nanobiotechnology
AlbaNova University Centre
Royal Institute of Technology
Stockholm, Sweden

ISSN 1064-3745 e-ISSN 1940-6029
ISBN 978-1-4939-5921-1 ISBN 978-1-61779-567-1 (eBook)
DOI 10.1007/978-1-61779-567-1
Springer New York Heidelberg Dordrecht London

Softcover re-print of the Hardcover 1st edition 2012

Printed on acid-free paper

Humana Press is part of Springer Science+Business Media (www.springer.com)

METHODS IN MOLECULAR BIOLOGY™

For further volumes:
http://www.springer.com/series/7651

Preface

Powerful tools for detailed cellular studies are emerging, increasing our knowledge of the ultimate target of all drugs: the living cell. Today, cells are commonly analyzed en masse, with thousands of cells per sample, yielding results on the average response of the cells. However, cellular heterogeneity implies that in order to learn more about cellular behavior, it is important to study how individual cells respond, one by one.

There are many conventional methods available for the study of single cells, such as flow cytometry, microscopy, capillary electrophoresis, and laser capture microdissection. These methods are typically robust, and their use has taught us almost everything we know about single-cell behavior in different environments. Researchers within the field of Life Science are continuously developing exciting new methods for analyzing entire cells on the proteomic, genomic, and transcriptomic level. Recently, miniaturized solutions have emerged as exciting and complementary ways of analyzing individual cells. With these microdevices and smaller volumes, experimental platforms are moving closer and closer to the size and volume of a single cell. Miniaturization can, for example, enable: (1) limited dilution of the cell content, (2) simultaneous analysis of several samples, and (3) an increase in experimental control by compartmentalization of single cells in droplets or wells. The methods described in this book include a few examples of conventional methods and several examples of miniaturized methods.

This book aims to serve as an update on the field of single-cell analysis, where the latest findings and applications are described in detail. The book is mainly not a review (although some review chapters are included); instead, it offers detailed protocols on how to carry out the methods in the laboratory. The invaluable "notes section" of each protocol chapter includes the kinds of tips and tricks that are crucial for succeeding with the experiments and yet are often left out of scientific publications. We believe that miniaturized systems are highly suitable for analyzing individual cells, but we are also aware that they often demand great laboratory skills and sometimes are far from "plug'n'play" systems that can be easily run by any end-user. By providing the readers with in-depth laboratory protocols, we hope that these methods can be made much more accessible and user-friendly. The book contains chapters written by leaders in the field and provides a firm background for anyone working with single-cell analysis. We hope that you will enjoy reading this book, which we believe is a very timely volume in the *Methods in Molecular Biology* series.

Below follows a short introduction to the book. The first chapter gives an introduction to single-cell analysis and discusses the huge progress and interest in the field. The three subsequent chapters deal with conventional techniques; Chap. 2 is a short review, introducing flow cytometry and microscopy as methods for analyzing single cells, Chap. 3 describes how peptide products from single cells can be identified using capillary electrophoresis and liquid chromatography-mass spectrometry, and Chap. 4 explains how laser capture microdissection can be used for gene expression analysis in single blastocysts.

The second part of the book deals with miniaturized solutions for single-cell analysis. The fifth chapter describes a method for culturing single cells into clones in order to study their heterogeneity in terms of proliferation. Chapter 6 demonstrates a method for encoding the cell's electrical properties, and can be used to identify cells that are electrically distinct from a background population. The protocol in Chap. 7 describes how individual cells are trapped in micrometer-size structures within a microchip, exposed in situ to a high electric field and loaded with plasmids. In Chap. 8, we learn more about how to resemble cellular microenvironments by trapping and culturing cells in a microfluidic channel with a build-in pillar array. The ninth chapter describes how padlock probes and rolling circle amplification can be used to detect DNA sequences within comet preparations, and in Chap. 10, we learn more about ways to analyze single cells using high-throughput droplet technologies. Chapter 11 demonstrates a method for screening antigen-specific antibody secreting cells using microwell-array chips. The twelfth chapter provides a generic protocol for the analysis of fixed and living single eukaryotic cells, including the considerations required to build a Raman Tweezers system. Chapter 13 describes a method for single-cell microinjection on nonadherent and adherent cells. In Chap. 14, we learn about how ultrasound can be combined with microfluidics and microplates for particle and cell manipulation approaching the single-cell level.

The third and final part of the book contains two review chapters. In Chap. 15, we can read about a rapidly growing field and emerging promising technologies that now enable sensitive "single-cell-omics" analysis. Lastly, Chap. 16 presents an engineering approach for integrated single-cell analysis that uses interchangeable modular operations to provide a comprehensive characterization of the phenotypic, functional, and genetic variations for individual cells.

Finally, we wish to thank all the authors that have contributed to this book; it is thanks to them that a volume on single-cell analysis could be realized. They have kindly shared their expertise by describing the exciting methodologies that they are working on, giving others a chance to fully understand and to try the methods themselves. In particular, we wish to thank them for generously, including all their "in-house expertise" in the Notes sections, containing the many tips and tricks that are needed in order to get the most out of the described assays. To the readers, if you have questions or comments that are not covered in the book, please do not hesitate to contact us for complementary information. We hope that you will enjoy reading this book and find it useful to learn more about the methods described and the available assays for single-cell analysis. Our aim is to encourage you to explore new ways of studying cells and help lead you to new discoveries.

Stockholm, Sweden

Sara Lindström, Ph.D.
Helene Andersson-Svahn, Ph.D.

Contents

Contributors

HELENE ANDERSSON-SVAHN • *Division of Nanobiotechnology, AlbaNova University Centre, Royal Institute of Technology, Stockholm, Sweden*

JULIE AUDET • *Institute of Biomaterials and Biomedical Engineering (IBBME), University of Toronto, Toronto, ON, Canada*

ALBERT VAN DEN BERG • *BIOS, The Lab-on-a-Chip Group, MESA+ Institute for Nanotechnology, University of Twente, Enschede, The Netherlands*

ERIC BROUZES • *Department of Biomedical Engineering, Stony Brook University, Stony Brook, NY, USA*

ROBERT B. BROWN • *Institute of Biomaterials and Biomedical Engineering (IBBME), University of Toronto, Toronto, ON, Canada*

DINO DI CARLO • *Department of Bioengineering, University of California, Los Angeles, CA, USA*

ANDREW EMILI • *Banting and Best Department of Medical Research, University of Toronto, Toronto, ON, Canada*

ELSA CORREIA FARIA • *Manchester Interdisciplinary Biocentre, School of Chemical Engineering and Analytical Science, The University of Manchester, Manchester, UK*

MURIEL FILLIERS • *Department of Reproduction, Obstetrics, and Herd Health, Faculty of Veterinary Medicine, Ghent University, Merelbeke, Belgium*

DOMINIQUE FOURMY • *LIMMS/CNRS-IIS, Institute of Industrial Science, University of Tokyo, Tokyo, Japan*

TERUO FUJII • *LIMMS/CNRS-IIS, JST-CREST, Institute of Industrial Science, University of Tokyo, Tokyo, Japan*

SÉVERINE LE GAC • *BIOS the Lab-on-a-Chip Group, MESA+ Institute for Nanotechnology, University of Twente, Enschede, The Netherlands*

PETER GARDNER • *Manchester Interdisciplinary Biocentre, School of Chemical Engineering and Analytical Science, The University of Manchester, Manchester, UK*

DANIEL R. GOSSETT • *Department of Bioengineering, University of California, Los Angeles, CA, USA*

SARA HENRIKSSON • *Department of Genetics and Pathology, Uppsala University, Uppsala, Sweden*

JOHANNES A. HEWEL • *Banting and Best Department of Medical Research, University of Toronto, Toronto, ON, Canada*

AISHUN JIN • *Department of Immunology, College of Basic Medical Science, Harbin Medical University, Nangang, Harbin, China*

SOO HYEON KIM • *JST-CREST, Institute of Industrial Science, University of Tokyo, Tokyo, Japan*

HIROYUKI KISHI • *Department of Immunology, Graduate School of Medicine and Pharmaceutical Sciences, University of Toyama, Toyama, Japan*

SARA LINDSTRÖM • *Department of Cell and Molecular Biology, Science For Life Laboratory, Karolinska Institute, Stockholm, Sweden*
J. CHRISTOPHER LOVE • *Department of Chemical Engineering, Koch Institute for Integrative Cancer Research, Massachusetts Institute of Technology, Cambridge, MA, USA*
ATSUSHI MURAGUCHI • *Department of Immunology, Graduate School of Medicine and Pharmaceutical Sciences, University of Toyama, Toyama, Japan*
MATS NILSSON • *Department of Genetics and Pathology, Uppsala University, Uppsala, Sweden*
TSUTOMU OBATA • *Central Research Institute, Toyama Industrial Technology Center, Toyama, Japan*
BJÖRN ÖNFELT • *Department of Microbiology, Tumour and Cell Biology, Karolinska Institute, Stockholm, Sweden*
TATSUHIKO OZAWA • *Department of Immunology, Graduate School of Medicine and Pharmaceutical Sciences, University of Toyama, Toyama, Japan*
ANN VAN SOOM • *Department of Obstretics, Reproduction and Herd Health, Faculty of Veterinary Medicine, Ghent University, Merelbeke, Belgium*
WARD DE SPIEGELAERE • *Department of Morphology, Faculty of Veterinary Medicine, Ghent University, Merelbeke, Belgium; Department of General Internal Medicine, Infectious Diseases & Psychosomatic Medicine, Faculty of Medicine and Health Sciences Ghent University Hospital, Ghent, Belgium*
KAZUTO TAJIRI • *The Third Department of Internal Medicine, Graduate School of Medicine and Pharmaceutical Sciences, University of Toyama, Toyama, Japan*
HENRY TAT KWONG TSE • *Department of Bioengineering, University of California, Los Angeles, CA, USA*
MICHAEL D. VAHEY • *Department of Electrical Engineering and Computer Science, Massachusetts Institute of Technology, Cambridge, MA, USA*
JOEL VOLDMAN • *Department of Electrical Engineering and Computer Science, Massachusetts Institute of Technology, Cambridge, MA, USA*
MARTIN WIKLUND • *Department of Biomedical and X-Ray Physics, Royal Institute of Technology, Stockholm, Sweden*
AYÇA YALÇIN • *Department of Chemical Engineering, Massachusetts Institute of Technology, Cambridge, MA, USA*
YVONNE J. YAMANAKA • *Department of Biological Engineering, Massachusetts Institute of Technology, Cambridge, MA, USA*
CHI ZHANG • *Division of Nanobiotechnology, AlbaNova University Centre, Royal Institute of Technology, Stockholm, Sweden*
YAN ZHANG • *Laboratory of Neurobiology and State Key Laboratory of Biomembrane and Membrane Biotechnology, School of Life Sciences, Peking University, Beijing, China*

Chapter 1

Introduction: Why Analyze Single Cells?

Dino Di Carlo, Henry Tat Kwong Tse, and Daniel R. Gossett

Abstract

Powerful methods in molecular biology are abundant; however, in many fields including hematology, stem cell biology, tissue engineering, and cancer biology, data from tools and assays that analyze the average signals from many cells may not yield the desired result because the cells of interest may be in the minority—their behavior masked by the majority—or because the dynamics of the populations of interest are offset in time. Accurate characterization of samples with high cellular heterogeneity may only be achieved by analyzing single cells. In this chapter, we discuss the rationale for performing analyses on individual cells in more depth, cover the fields of study in which single-cell behavior is yielding new insights into biological and clinical questions, and speculate on how single-cell analysis will be critical in the future.

Key words: Single-cell analysis, Microfluidics, Biomicrofluidics, Cellular heterogeneity

The development of new technology is one of the principal driving forces of biological discovery. Tools that assay the amount and type of biomolecules within cells, such as real-time PCR and high-throughput DNA sequencers, have enabled us to approach new problems and led to the emergence of new fields. Especially now, the availability and breadth of molecular and cellular assays is growing rapidly—with increasingly refined information obtainable about the levels of a myriad of biomolecules within cells. Powerful methods in molecular biology are abundant. However, in many fields including hematology, stem cell biology, tissue engineering, and cancer biology, data from tools and assays that analyze the average response from many cells is difficult to interpret. Some of this difficulty is attributed to population heterogeneity and varied time dynamics within the sample, which makes measures of the whole cell population misleading and necessitates molecular biology methods with single-cell resolution. The development of single-cell analysis tools such as flow cytometers has been spurred, in part, by the complexities of the cell samples in fields such as hematology. Here, we highlight this and other rationale for performing analyses on individual cells in more detail. We introduce emerging fields of

Sara Lindström and Helene Andersson-Svahn (eds.), *Single-Cell Analysis: Methods and Protocols*,
Methods in Molecular Biology, vol. 853, DOI 10.1007/978-1-61779-567-1_1, © Springer Science+Business Media, LLC 2012

study in which single-cell behavior is yielding new insights into biological and clinical questions. We also provide an overview of new approaches, discussed in more detail throughout the book, that are being developed to address these emerging fields and expand the capabilities of single-cell analysis.

Heterogeneity within a population of cells often necessitates the careful interpretation of results; traditional assays yield blended levels of molecules within the population. Considering the ease of data collection, it is more important than ever to collect meaningful information—to have the right tool for the job. Oftentimes, tools that report on molecular level information of an ensemble of cells are not well-suited to study complex, heterogeneous samples like those found in stem-cell biology, immunology, and tissue engineering. Consider how these molecular assays are often performed. All cells are stimulated uniformly then analyzed with bulk measurement techniques (e.g., western blots, to determine an average cellular response in terms of biomolecule content). These conventional methods are accessible and simple but can incorrectly describe the distribution of behavior among cells in the sample. For example, single-cell time-responses can be masked in bulk measures. This is important because even genetically identical cells are dynamic and varied in their responses to stimuli due to epigenetic differences, the stochastic nature of intracellular signaling, and the small amount of molecules which are engaged (1, 2). Further, often cells of interest (in tumors or complex tissues) are among populations of genetically different cells. In both of these cases, particular cells of interest may respond strongly to stimuli, but other subgroups in the sample may respond weakly. Measured batch kinetics will not be representative of the actual response at the single-cell level (Fig. 1a). Now, consider a distribution of gene expression levels within a population that is not normally distributed. Indeed examples of bimodal expressions exist as cells respond dynamically to extracellular conditions (3). A bulk measurement of this population will output a mean value that is not representative of either subpopulation (Fig. 1b–d). Only by measuring the properties of single cells will the true heterogeneity and behavior of the population be observed.

Awareness of population heterogeneity has lead to valuable insights in the fields of systems biology, stem cell biology, cancer biology, and tissue engineering. These areas have benefited greatly from the emergence of single-cell analysis methods (Fig. 2), using new tools such as gene expression profile mapping and methods for sorting complex samples (4). In cancer biology, the heterogeneous nature of solid tumors has long been known (5, 6) and is of immense interest due to the possibility of inferring disease progression from genomic heterogeneity and gaining insights into the roles and impact of cancer stem cells (7, 8). Using single-cell resolution techniques, dynamic proteomics has been used to track population subgroups to identify initial therapeutic potential of multiple drugs

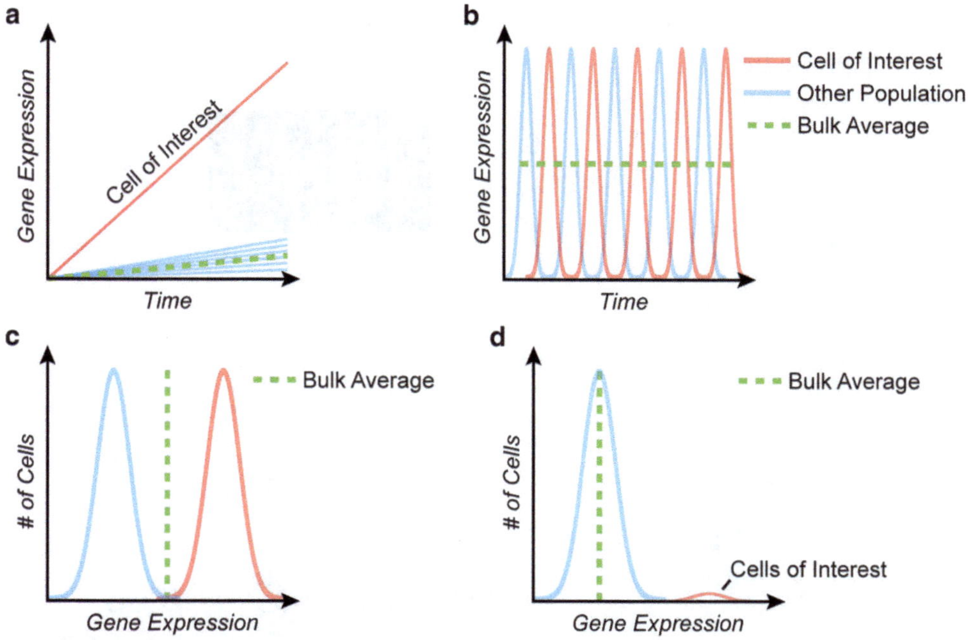

Fig. 1. Heterogeneity of cellular behavior, with gene expression as an example. (**a**) Bulk measurement of a rate masks amplitude differences between cells of interest and the majority population and (**b**) out-of-phase dynamics. (**c**) Subpopulations in multimodal distributions and (**d**) rare cell populations are not accurately represented by ensemble end-point assays.

and the rise in resistance (9, 10). Further, single-cell gene expression techniques have been used to analyze the regulatory relationships that exist early in development and in the differentiation of stem cells. Conventionally, bulk RNA analysis has been used to study gene regulation. It was assumed human embryonic stem cells are genetically homogeneous. However, using single-cell transcriptome analysis, highly heterogeneous genetic expression profiles existing in pluripotent human embryonic stem cell colonies have been observed, suggesting a need for better characterization and control of these initial populations (11)—something bulk methods lacked the sensitivity to achieve.

Microfabrication techniques have enabled miniaturized systems which expand upon the capabilities of current tools: systems which perform conventional assays on single cells as well as tools which can assist in controlling microenvironments or separating cells to extract pure populations. Conventional analysis methods have been miniaturized and new analysis tools unique to these platforms have been developed. Microfluidic lab-on-a-chip (LOC) platforms have been developed to interface with and assay single cells. Some of these tools offer precise control of mechanical and chemical environmental cues (12–15), and the collection of highly quantifiable—often temporal—single-cell data points from various assays (16–18). The acquisition of single-cell dynamics will prove especially useful for identifying rare cells which exhibit unique dynamic behavior

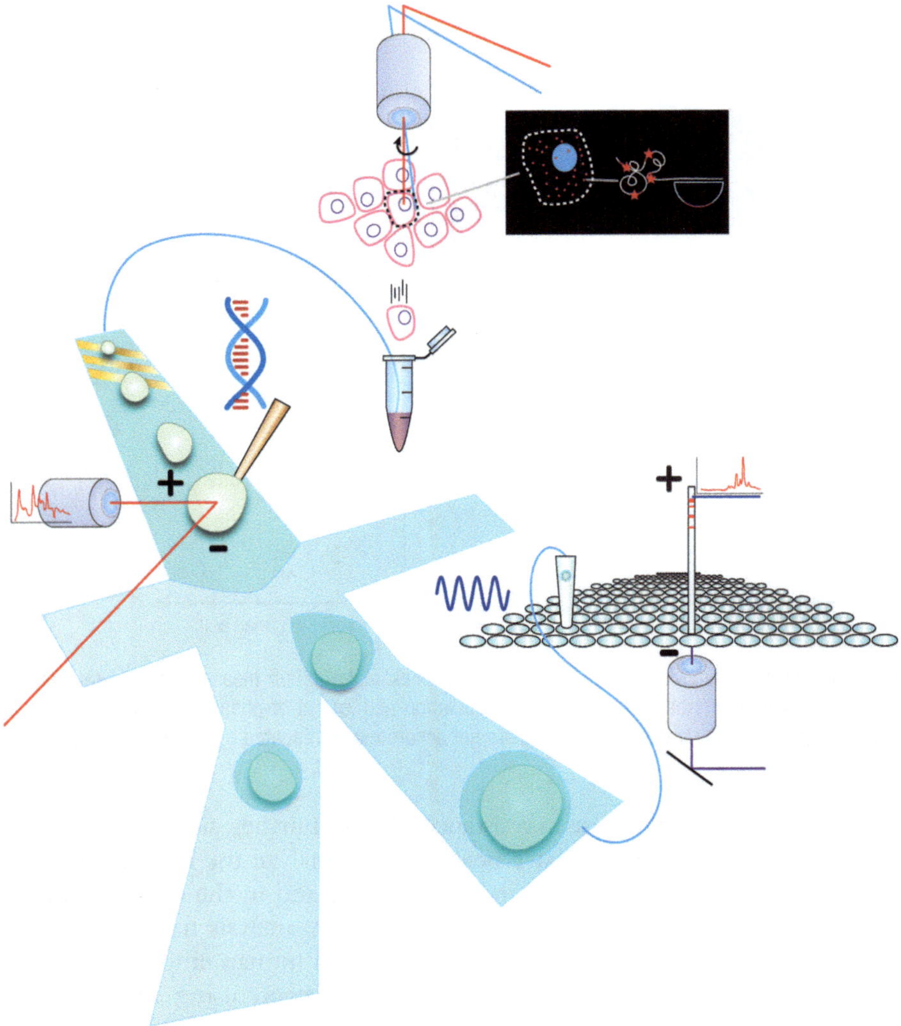

Fig. 2. Single-cell analysis tools and methods.

such as a high drug efflux rate, fast migration, high proliferation, or unique cytokinetic behavior. Often, these miniaturized systems incorporate cell culturing techniques prior to end-point analysis to evaluate cell–cell interactions (19, 20), cellular response to substrates (21), and coculture communication (22), and to study biomimetic applications (23, 24). Miniaturized systems can better mimic in vivo microscale environments such as biochemical gradients on the level of the single cell, which is not achievable with conventional macroscale tissue culture methods. Observation of single cells in these microenvironments enables the study of unique phenomena such as cancer-cell-induced angiogenesis (25). For analysis of more complex samples like blood (blood cell counting or rare cell detection and sorting, sorting antigen-specific T cells) a high-throughput is required. The engineering of high-throughput single-cell analysis tools has required the development of methods to manipulate or handle many cells: assigning precise positions in arrays or positioning in flow. Modes of manipulation such as physical trapping (26), inertial focusing/separation (27), dielectric focusing/separation (28), and acoustic focusing/separation (29) are derived from phenomena which can only be accessed practically at the microscale. Understanding the advantages rendered by miniaturized tools has led to investigations of stem cell colony heterogeneity (single-cell gene expression), new methods of biochemical detection and analysis (biochemical immunoassays) (30), and biophysical effects such as shear stress on adhesion dynamics (31). Further, some systems incorporate cell culture with miniaturized end-point assays such as PCR (32) and ELISA (33, 34). These systems have the potential to impact both research and biomedicine by providing information-rich datasets derived from single-cell manipulation and measurements.

Beyond the ability to interface at the level of individual cells, miniaturization confers a number of advantages including the ability to increase throughput via parallelization and reduce reagent costs and human error via automation (35). A number of the miniaturized systems and methods described in this text take advantage of microfabrication, sample automation, and continuous flow to parallelize conventional single-cell assays or perform them serially more rapidly. Microfabricated tools such as microwell and microchamber arrays increase throughput by miniaturization and parallelization. While, the number of cells assayed with these techniques (up to 234,000 cells in the smallest wells (36)) may not compare to flow cytometry, flow cytometry is less useful for studying dynamics, especially dynamics on short timescales (36, 37). In microwells, on the other hand, cell signaling and response to stimuli can be tracked. In fact, there are endless applications of these new methods, from massively parallel, single-cell/clone, long-term culture (38) to PCR (39) to detecting and obtaining secreted antibodies from single cells for immune therapy (36). Further, the small volume of the microwells minimizes reagent volume, reducing the cost of the assay. A complementary method, also described in this text, is ultrasonic manipulation of single cells. This technique can be applied to multiwell plates, driving the aggregation of cells to known locations within the wells. This technique could be useful to coordinate the initiation of interactions between different cell types for the purpose of studying the dynamics of their behavior (40).

Samples where the cells of interest are rare *demand* single-cell measurements or separation methods for achieving high levels of subpopulation purity to achieve necessary sensitivity. It is instructive to consider assays of rare cells in blood: hematopoietic cells, antigen specific T cells, circulating fetal cells in maternal blood, or circulating tumor cells. Within a milliliter of blood there are billions of cells but the population of interests may range from only tens to thousands. Even with the gold-standard, commercial, single-cell analysis tool, such as the flow cytometer, which can measure multiple cell properties at a high-throughput (up to 100,000 cells/s), it takes hours to identify these rare events (41). Despite the critical importance of these applications this lengthy occupation of core equipment is impractical. Therefore, sample preparation (e.g., positive or negative selection) is often a necessary phase in single-cell analysis.

A number of bulk and single-cell tools exist to sort cells, such as fluorescence activated cell sorting (FACS) and magnetic activated cell sorting (MACS), but these technologies rely on costly antibody labeling. Often, the rare cells of interest have distinct physical attributes relative to the populous which are exploited by methods found in this text. Measurement of and manipulation based on these intrinsic biomarkers is a strategy to increase purity or perform label-free single-cell measurements at a low cost. These can be cytometric assays or sample preparation techniques, but in principle, the methods should rapidly process single cells in series. Iso-dielectric separation (IDS) is an example of both. As a label-free method of cell sorting (42), IDS, manipulates an intrinsic property (electrical properties and size) of cells to sort heterogeneous populations without the use of biochemical, fluorescent tags. Thus, inherent in the sample preparation or sorting is the analysis of cell electrical properties, which can vary within a sample (e.g., live and dead cells or the components of blood) (43, 44). Another example of a label-free assay is the use of Raman tweezers to classify single cells based on their biochemical spectra. Differences in Raman spectra exist in many relevant comparisons of eukaryotic cells—for example, between prostate cancer cell lines and bladder cell lines. A noninvasive assay which could distinguish prostate cancer cells from bladder cells and other benign cells in the urine would be potentially more powerful than convention, which is a bulk measurement of prostate-specific antigen (PSA) in the blood. The bulk measurement yields many false positives because PSA levels can increase due to other conditions, such as the presence of benign tumors or inflammation (45, 46).

Realization of the value of single-cell analysis has led to the development of many single-cell molecular biology techniques, even outside the realm of miniaturized systems. A sample of methods for single-cell analysis which are covered in this text include single-cell capillary electrophoresis (CE) or mass spectrometry for protein quantification and separation, and laser capture microdissection (LCM) for gene expression analysis of specific cell populations in single blastocysts. Not only is single-cell CE important for studying heterogeneous populations like those in cancer and blood, it has also been shown to reduce artifacts from nonmetabolic processes and bias present in measurement of the extracts of several cells (47, 48). Another complex tissue is the blastocyst which is of critical interest in developmental biology. While a typical isolation may involve digestion or disaggregation of the tissue prior to separation, LCM allows for isolation of specific subpopulations without destruction of the microenvironment or possible contamination by other cells during chemical treatments. This makes it an especially useful method for assaying single-cell gene expression and molecular processes (49, 50).

To supplant or coexist with gold standard molecular biology methods, single-cell methods will not only need to provide access to previously immeasurable phenomena but also be easy to use and robust. Typically, this means multistep processes are streamlined and packaged such that the method can be adopted without much technical skill. Droplet-based analysis of single cells (51) has matured to the point where the modular components: droplet generation, reactant loading, and droplet manipulation have been integrated on a single platform. The aim of this text is to make single-cell analyses more accessible, and the methods described here possess these qualities, which should encourage their adoption in the lab.

For research applications, besides evaluating single-cell response to externally applied chemicals or stimuli, investigation of intracellular stimuli is of interest in understanding cellular function. Bulk methods of delivering cargo into cells can be chemical or electrical and typically lacks the requisite uniformity and efficiency. Due to these deficiencies, it is impossible to directly attribute differences in single-cell response to population heterogeneity. Further, these methods are toxic to certain types of cells. Presented here are two methods for delivering cargo to single cells: microinjection and single-cell electroporation (52–54). While electroporation is parallelized in a miniaturized system, microinjection remains serial; however, it has been proposed that micropatterned substrates should be used to position and orient cells for automated microinjection to enhance throughput (52). The application of the methods will depend on the type of cells and cargo, and increased multiplexing will enable the use of the methods for problems with magnified complexity and rarity. One especially challenging problem is detecting rare transcript variants. Bulk methods cannot measure these molecules when the cells of interest are exceedingly rare. Further, bulk methods like PCR and Western blots, which require large quantities of biomolecules, do not perform well when the molecules of interest are rare. One conventional approach to this problem is LCM followed by single-cell PCR; however, this process is time consuming, error prone, and generally not well suited to diagnostics. A method in this text describes how these variants can be identified in single cells using padlock probes and target-primed rolling-circle amplification (55). This assembly of useful methods and reviews should provide a foundation for biologists to begin analyzing single cells when doing so provides more accurate detection or measurement.

The tools discussed here and other single-cell analysis techniques should not be restrictive to research laboratories. While genetic (56) and proteomic (57) analyses are currently available, they are rarely performed in the health care setting due to cost-inefficiencies, the technical complexity of the assays, and the heterogeneity of the tissue samples. However, automated, miniaturized systems are potentially cost-effective (minimized labor and reagent cost), easier to use, and enable access to single cells and critical disease causing subpopulations. The ability to collect information rich, single-cell-based, molecular datasets should lead to a personalized medicine approach to disease diagnosis and treatment. For example, the treatment of highly heterogeneous tumor tissues can be improved by analysis of biopsies at the molecular level with single-cell resolution for expression of known drug-resistant efflux pumps. This data can then provide physicians a better suggestion of appropriate therapeutic regimens. Quantitative single-cell methods coupled with advances in medical information systems will lead to more accurate decisions by care-providers leading to shorter patient stays in hospitals and fewer drug regimens, all helping to ease the cost of health care.

References

1. Elowitz MB, Levine AJ, Siggia ED & Swain PS (2002) Stochastic Gene Expression in a Single Cell. *Science* **297**, 1183–1186.
2. Miller MJ, Safrina O, Parker I & Cahalan MD (2004) Imaging the Single Cell Dynamics of CD4+ T Cell Activation by Dendritic Cells in Lymph Nodes. *The Journal of Experimental Medicine* **200**, 847–856.
3. Fiering S, Northrop JP, Nolan GP, Mattila PS, Crabtree GR & Herzenberg LA (1990) Single cell assay of a transcription factor reveals a threshold in transcription activated by signals emanating from the T-cell antigen receptor. *Genes & Development* **4**, 1823–1834.
4. Levsky JM, Shenoy SM, Pezo RC & Singer RH (2002) Single-Cell Gene Expression Profiling. *Science* **297**, 836–840.
5. Dexter DL, Kowalski HM, Blazar BA, Fligiel Z, Vogel R & Heppner GH (1978) Heterogeneity of Tumor Cells from a Single Mouse Mammary Tumor. *Cancer Research* **38**, 3174–3181.
6. Dexter DL, Spremulli EN, Fligiel Z, Barbosa JA, Vogel R, VanVoorhees A & Calabresi P (1981) Heterogeneity of cancer cells from a single human colon carcinoma. *Am. J. Med* **71**, 949–956.
7. Vermeulen L, Sprick MR, Kemper K, Stassi G & Medema JP (2008) Cancer stem cells - old concepts, new insights. *Cell Death Differ* **15**, 947–958.
8. Navin N, Krasnitz A, Rodgers L, Cook K, Meth J, Kendall J, Riggs M, Eberling Y, Troge J, Grubor V, Levy D, Lundin P, Månér S, Zetterberg A, Hicks J & Wigler M (2010) Inferring tumor progression from genomic heterogeneity. *Genome Research* **20**, 68–80.
9. Orth JD, Tang Y, Shi J, Loy CT, Amendt C, Wilm C, Zenke FT & Mitchison TJ (2008) Quantitative live imaging of cancer and normal cells treated with Kinesin-5 inhibitors indicates significant differences in phenotypic responses and cell fate. *Molecular Cancer Therapeutics* **7**, 3480–3489.
10. Cohen AA, Geva-Zatorsky N, Eden E, Frenkel-Morgenstern M, Issaeva I, Sigal A, Milo R, Cohen-Saidon C, Liron Y, Kam Z, Cohen L, Danon T, Perzov N & Alon U (2008) Dynamic Proteomics of Individual Cancer Cells in Response to a Drug. *Science* **322**, 1511–1516.
11. Zhong JF, Chen Y, Marcus JS, Scherer A, Quake SR, Taylor CR & Weiner LP (2008) A microfluidic processor for gene expression profiling of single human embryonic stem cells. *Lab Chip* **8**, 68.
12. Kim L, Vahey MD, Lee H & Voldman J (2006) Microfluidic arrays for logarithmically perfused embryonic stem cell culture. *Lab Chip* **6**, 394–406.
13. Vermeulen L, Todaro M, de Sousa Mello F, Sprick MR, Kemper K, Perez Alea M, Richel DJ, Stassi G & Medema JP (2008) Single-cell cloning of colon cancer stem cells reveals a multi-lineage differentiation capacity. *Proceedings of the National Academy of Sciences* **105**, 13427–13432.

14. Chung BG, Flanagan LA, Rhee SW, Schwartz PH, Lee AP, Monuki ES & Jeon NL (2005) Human neural stem cell growth and differentiation in a gradient-generating microfluidic device. *Lab Chip* **5**, 401.
15. Weaver WM, Dharmaraja S, Milisavljevic V & Di Carlo D (2011) The effects of shear stress on isolated receptor–ligand interactions of Staphylococcus epidermidis and human plasma fibrinogen using molecularly patterned microfluidics. *Lab Chip* **11**, 883.
16. Di Carlo D, Aghdam N & Lee LP (2006) Single-Cell Enzyme Concentrations, Kinetics, and Inhibition Analysis Using High-Density Hydrodynamic Cell Isolation Arrays. *Analytical Chemistry* **78**, 4925–4930.
17. Di Carlo, D & Lee LP (2006) Dynamic Single-Cell Analysis for Quantitative Biology. *Analytical Chemistry* **78**, 7918–7925.
18. Gossett DR, Weaver WM, Ahmed NS & Di Carlo D (2010) Sequential Array Cytometry: Multi-Parameter Imaging with a Single Fluorescent Channel. *Ann Biomed Eng* **39**, 1328–1334.
19. Wright D, Rajalingam B, Selvarasah S, Dokmeci MR & Khademhosseini A (2007) Generation of static and dynamic patterned co-cultures using microfabricated parylene-C stencils. *Lab Chip* **7**, 1272–1279.
20. Chung S, Sudo R, Mack PJ, Wan C, Vickerman V & Kamm RD (2009) Cell migration into scaffolds under co-culture conditions in a microfluidic platform. *Lab Chip* **9**, 269–275.
21. Ochsner M, Dusseiller MR, Grandin HM, Luna-Morris S, Textor M, Vogel V & Smith ML (2007) Micro-well arrays for 3D shape control and high resolution analysis of single cells. *Lab Chip* **7**, 1074.
22. Hui EE & Bhatia SN (2007) Micromechanical control of cell–cell interactions. *Proceedings of the National Academy of Sciences* **104**, 5722–5726.
23. Toh Y, Ng S, Khong YM, Samper V & Yu H (2005) A configurable three-dimensional microenvironment in a microfluidic channel for primary hepatocyte culture. *Assay Drug Dev Technol* **3**, 169–176.
24. Zhang MY, Lee PJ, Hung PJ, Johnson T, Lee LP & Mofrad MRK (2007) Microfluidic environment for high density hepatocyte culture. *Biomed Microdevices* **10**, 117–121.
25. Chung S, Sudo R, Mack PJ, Wan C, Vickerman V & Kamm RD (2009) Cell migration into scaffolds under co-culture conditions in a microfluidic platform. *Lab Chip* **9**, 269.
26. Rettig JR & Folch A (2005) Large-Scale Single-Cell Trapping And Imaging Using Microwell Arrays. *Analytical Chemistry* 77, 5628–5634.
27. Di Carlo D, Edd JF, Irimia D, Tompkins RG & Toner M (2008) Equilibrium separation and filtration of particles using differential inertial focusing. *Anal. Chem* **80**, 2204–2211.
28. Wang X, Yang J, Huang Y, Vykoukal J, Becker FF & Gascoyne PRC (2000) Cell Separation by Dielectrophoretic Field-flow-fractionation. *Analytical Chemistry* **72**, 832–839.
29. Evander M, Johansson L, Lilliehorn T, Piskur J, Lindvall M, Johansson S, Almqvist M, Laurell T & Nilsson J (2007) Noninvasive Acoustic Cell Trapping in a Microfluidic Perfusion System for Online Bioassays. *Analytical Chemistry* **79**, 2984–2991.
30. Choi J, Oh KW, Thomas JH, Heineman WR, Halsall HB, Nevin JH, Helmicki AJ, Henderson HT & Ahn CH (2002) An integrated microfluidic biochemical detection system for protein analysis with magnetic bead-based sampling capabilities. *Lab Chip* **2**, 27.
31. Lu H, Koo LY, Wang WM, Lauffenburger DA, Griffith LG & Jensen KF (2004) Microfluidic Shear Devices for Quantitative Analysis of Cell Adhesion. *Analytical Chemistry* **76**, 5257–5264.
32. Khandurina J, McKnight TE, Jacobson SC, Waters LC, Foote RS & Ramsey JM (2000) Integrated System for Rapid PCR-Based DNA Analysis in Microfluidic Devices. *Analytical Chemistry* **72**, 2995–3000.
33. Sato K, Yamanaka M, Takahashi H, Tokeshi M, Kimura H & Kitamori T (2002) Microchip-based immunoassay system with branching multichannels for simultaneous determination of interferon-gamma. *Electrophoresis* **23**, 734–739.
34. Sato K, Tokeshi M, Odake T, Kimura H, Ooi T, Nakao M & Kitamori T (2000) Integration of an immunosorbent assay system: analysis of secretory human immunoglobulin A on polystyrene beads in a microchip. *Anal. Chem* **72**, 1144–1147.
35. Zare RN & Kim S (2010) Microfluidic platforms for single-cell analysis. *Annu Rev Biomed Eng* **12**, 187–201.
36. Jin A, Ozawa T, Tajiri K, Obata T, Kondo S, Kinoshita K, Kadowaki S, Takahashi K, Sugiyama T, Kishi H & Muraguchi A (2009) A rapid and efficient single-cell manipulation method for screening antigen-specific antibody-secreting cells from human peripheral blood. *Nat Med* **15**, 1088–1092.
37. Yamamura S, Kishi H, Tokimitsu Y, Kondo S, Honda R, Rao SR, Omori M, Tamiya E & Muraguchi A (2005) Single-Cell Microarray for Analyzing Cellular Response. *Analytical Chemistry* 77, 8050–8056.

38. Lindström S, Larsson R & Andersson Svahn H (2008) Towards high-throughput single cell/clone cultivation and analysis. *Electrophoresis* **29**, 1219–1227.
39. Lindström S, Hammond M, Brismar H, Andersson-Svahn H & Ahmadian A (2009) PCR amplification and genetic analysis in a microwell cell culturing chip. *Lab Chip* **9**, 3465.
40. Vanherberghen B, Manneberg O, Christakou A, Frisk T, Ohlin M, Hertz HM, Önfelt B & Wiklund M (2010) Ultrasound-controlled cell aggregation in a multi-well chip. *Lab Chip* **10**, 2727.
41. Shapiro HM (2003) *Practical Flow Cytometry*, 4th ed. Wiley-Liss, New York.
42. Gossett DR, Weaver WM, Mach AJ, Hur SC, Tse HTK, Lee W, Amini H & Di Carlo D (2010) Label-free cell separation and sorting in microfluidic systems. *Anal Bioanal Chem* **397**, 3249–3267.
43. Vahey MD & Voldman J (2009) High-Throughput Cell and Particle Characterization Using Isodielectric Separation. *Analytical Chemistry* **81**, 2446–2455.
44. Vahey MD & Voldman J (2008) An Equilibrium Method for Continuous-Flow Cell Sorting Using Dielectrophoresis. *Analytical Chemistry* **80**, 3135–3143.
45. Harvey TJ, Hughes C, Ward AD, Correia Faria E, Henderson A, Clarke NW, Brown MD, Snook RD & Gardner P (2009) Classification of fixed urological cells using Raman tweezers. *J Biophotonics* **2**, 47–69.
46. Snook RD, Harvey TJ, Correia Faria E & Gardner P (2009) Raman tweezers and their application to the study of singly trapped eukaryotic cells. *Integr. Biol.* **1**, 43.
47. Krylov SN, Arriaga E, Zhang Z, Chan NWC, Palcic MM & Dovichi NJ (2000) Single-cell analysis avoids sample processing bias. *Journal of Chromatography B: Biomedical Sciences and Applications* **741**, 31–35.
48. Brehm-Stecher BF & Johnson EA (2004) Single-Cell Microbiology: Tools, Technologies, and Applications. *Microbiol. Mol. Biol. Rev.* **68**, 538–559.
49. Espina V, Wulfkuhle JD, Calvert VS, VanMeter A, Zhou W, Coukos G, Geho DH, Petricoin EF & Liotta LA (2006) Laser-capture microdissection. *Nat. Protocols* **1**, 586–603.
50. Filliers M, De Spiegelaere W, Peelman L, Goossens K, Burvenich C, Vandaele L, Cornillie P & Van Soom A (2011) Laser capture microdissection for gene expression analysis of inner cell mass and trophectoderm from blastocysts. *Analytical Biochemistry* **408**, 169–171.
51. Brouzes E, Medkova M, Savenelli N, Marran D, Twardowski M, Hutchison JB, Rothberg JM, Link DR, Perrimon N & Samuels ML (2009) Droplet microfluidic technology for single-cell high-throughput screening. *Proc. Natl. Acad. Sci. USA* **106**, 14195–14200.
52. Zhang Y & Yu L (2008) Microinjection as a tool of mechanical delivery. *Curr. Opin. Biotechnol* **19**, 506–510.
53. Zhang Y & Yu L (2008) Single-cell microinjection technology in cell biology. *Bioessays* **30**, 606–610.
54. Valero A, Post JN, van Nieuwkasteele JW, ter Braak PM, Kruijer W & van den Berg A (2008) Gene transfer and protein dynamics in stem cells using single cell electroporation in a microfluidic device. *Lab Chip* **8**, 62.
55. Larsson C, Grundberg I, Soderberg O & Nilsson M (2010) In situ detection and genotyping of individual mRNA molecules. *Nat Meth* 7, 395–397.
56. Heidecker B & Hare JM (2007) The use of transcriptomic biomarkers for personalized medicine. *Heart Fail Rev* **12**, 1–11.
57. Liotta LA, Kohn EC & Petricoin EF (2001) Clinical Proteomics. *JAMA: The Journal of the American Medical Association* **286**, 2211–2214.

Part I

Conventional Methods

Chapter 2

Flow Cytometry and Microscopy as Means of Studying Single Cells: A Short Introductional Overview

Sara Lindström

Abstract

Flow cytometry and microscopy are perhaps the two most commonly used techniques for analyzing single cells. Both techniques are typically robust and provide a high throughput analysis of living and/or fixed cells. The techniques are often combined with fluorescent labeling of cells, using for example antibodies. This chapter is a short introductory review where some of the possible applications using flow cytometry and microscopy are discussed.

Key words: Flow cytometry, Microscopy, Single cells, High throughput

Single mammalian cells are fragile, small and have strict requirements of their microenvironment to maintain life (e.g., oxygen, temperature, nutrition, etc.). By nature, cell lines are more tolerant and thus easier to work with, compared to primary cells. When analyzing single cells, it is important to study many single cells simultaneously in order not to draw misleading or wrong conclusions from rare cells or stochastic biological noise. How many single cells must then be experimentally investigated to acquire adequate sets of data in order to safely draw conclusions? This question has (of course) no general answer since it depends on the application. Conventionally, 10,000 cells are considered a standard number (1), and it has been suggested that aiming for 1,000 cells might be enough when developing general single-cell technologies (2). However, the importance of analyzing a large number of individual cells and determining the distribution of responses, due to cell heterogeneity, is undisputed and has been highlighted previously (3, 4).

The most commonly used method for single-cell analysis is flow cytometry (FC), developed in the late 1960s. Flow cytometric technology origins from the three strands of microscopy, blood cell counting instruments, and ink jet technology developed for computer printers. FC allows hundreds of thousands of individual cells per minute to be analyzed according to their size, granularity, and

Sara Lindström and Helene Andersson-Svahn (eds.), *Single-Cell Analysis: Methods and Protocols*,
Methods in Molecular Biology, vol. 853, DOI 10.1007/978-1-61779-567-1_2, © Springer Science+Business Media, LLC 2012

fluorescence properties in a wide range of applications, e.g., viability, protein expression and localization, gene expression, etc. (5). Cell sorting by fluorescence (FACS, fluorescence-activated cell sorting) enables sorting one or several populations of cells from a mixed sample, for further analysis (the sorted cells are normally collected in test tubes or in multiwell plates). FC is a "state-of-the-art" technique for well-characterized distributions of individual cell behavior at high throughput (up to 10,000 cells/s). The reason for its success is the throughput combined with fluorescent labeling, allowing semiquantitative determination of for example various protein levels in a population of cells (6, 7). It is also possible to analyze cells in a time-dependent manner, where cells are sampled at different time points by FC (8). However, time-dependent studies on individual cells using FC are not commonly performed since the analyzed cells often go to waste or are sorted (i.e., often mixed) with other cells. Conventional FC require many cells for analysis (at least 100,000 cells) and cells must be mixed again (thereby lost track of) before a second round of analysis, hence FC gives information of the distribution of a group of cells. Using FC, it is not possible to follow single cells over time. Likewise, tracking of cell divisions using FC is performed in bulk (9), and FC was neither designed for handling, manipulation, and dynamic analysis of single cells nor observation of spatial localization of fluorescence within a cell.

Microscopy, another frequently used method for single-cell analysis, is well suited for intracellular localization- and time-dependent studies, and the field of live-cell imaging is built upon monitoring individual cell behavior. The use of microscopy in biological research was introduced in the mid 17th century and is today used on a daily basis in most biological laboratories. As discussed previously, the importance of studying large numbers of cells with accurate interpretation, have resulted in the field of automated microscopy (AM), also referred to as high-throughput microscopy (HTM), image cytometry (IC), screening, cellomics, imaging, high content screening/analysis (HCS/HCA) (10, 11). Normally, cells are either fixed or studied dynamically, and hundreds of images are obtained in a rapid manner followed by massive image analysis to extract useful information about the cells and their intracellular compartments. Substantial limitations of AM involve throughput, multiparametric assays, and time spent on data acquisition. AM is often used to study average cell behavior but also holds the potential for obtaining results on the heterogeneity of cell samples, though often lacking controlled seeding patterns and identification of cell boundaries for rapid analysis. It is often cumbersome to monitor and track cell divisions, since suitable methods for isolating single cells or clones have not been routinely used. The conventional tools (e.g., 96/384-well plates) used for AM are relatively big as compared to the size of a cell. More importantly, the wells of such a multiwell plate are unnecessarily large as compared to the area covered by the objective's "field of view" (even when using low magnification optics), sometimes making AM in combinations with conventional tools less suited for monitoring individual cells.

Examples of other conventional techniques for single-cell analysis are as follows: (1) Laser scanning cytometry (LSC), which allows imaging and quantitative analysis of individual cells in tissues in-situ (12), (2) Capillary electrophoresis (CE) for efficient separation and sensitive detection of whole cell or subcellular samples (13), and (3) Laser capture microdissection (LCM) for cutting out and separating single cells from tissue for further analysis, such as gene expression and protein analysis (14). As with many methods, the throughput is a constant challenge and the type of analyses that one can perform using the three techniques mentioned above, is thus somewhat restricted. There are several applications lacking good tools for single-cell analysis, for example investigating perforated or encapsulated cells, cellular behavior, and cell–cell interactions in microengineered environments, as reviewed previously (15). Additional novel applications are most certain to appear, as appropriate tools for robust single-cell analyses become increasingly available.

References

1. Givan A (2001) Flow cytometry first principles. *Wiley-Liss* second edition.
2. Andersson Svahn H, Van den Berg, A. (2007) Single cells or large populations? *Lab on a chip* 7(5):544–546.
3. Mettetal JT, Muzzey D, Pedraza JM, Ozbudak EM, & van Oudenaarden A (2006) Predicting stochastic gene expression dynamics in single cells. *Proceedings of the National Academy of Sciences of the United States of America* 103(19):7304–7309.
4. Yu J, Xiao J, Ren X, Lao K, & Xie XS (2006) Probing gene expression in live cells, one protein molecule at a time. *Science* 311(5767):1600–1603.
5. Villas BH (1998) Flow cytometry: an overview. *Cell Vis* 5(1):56–61.
6. Krutzik PO & Nolan GP (2006) Fluorescent cell barcoding in flow cytometry allows high-throughput drug screening and signaling profiling. *Nature methods* 3(5):361–368.
7. Nolan JP & Sklar LA (1998) The emergence of flow cytometry for sensitive, real-time measurements of molecular interactions. *Nature biotechnology* 16(7):633–638.
8. Martin JC & Swartzendruber DE (1980) Time: a new parameter for kinetic measurements in flow cytometry. *Science* 207(4427):199–201.
9. Roostalu J, Joers A, Luidalepp H, Kaldalu N, & Tenson T (2008) Cell division in Escherichia coli cultures monitored at single cell resolution. *BMC microbiology* 8:68.
10. Oheim M (2007) High-throughput microscopy must re-invent the microscope rather than speed up its functions. *British journal of pharmacology* 152(1):1–4.
11. Pepperkok R & Ellenberg J (2006) High-throughput fluorescence microscopy for systems biology. *Nat Rev Mol Cell Biol* 7(9):690–696.
12. Harnett MM (2007) Laser scanning cytometry: understanding the immune system in situ. *Nat Rev Immunol* 7(11):897–904.
13. Arcibal IG, Santillo MF, & Ewing AG (2007) Recent advances in capillary electrophoretic analysis of individual cells. *Anal Bioanal Chem* 387(1):51–57.
14. Kehr J (2003) Single cell technology. *Current opinion in plant biology* 6(6):617–621.
15. Sims CE & Allbritton NL (2007) Analysis of single mammalian cells on-chip. *Lab on a chip* 7(4):423–440.

Chapter 3

Identification of Enzyme-Converted Peptide Products from Single Cells Using Capillary Electrophoresis and Liquid Chromatography-Mass Spectrometry

Robert B. Brown, Johannes A. Hewel, Andrew Emili, and Julie Audet

Abstract

Single-cell analysis using chemical methods, otherwise known as chemical cytometry, promises to provide significant leaps in understanding signaling processes which result in cellular behavior. Sensitive methods for chemical cytometry such as capillary electrophoresis can detect and quantify multiple targets; however, conclusive identification of detected analytes is required for useful data to be obtained. Here, we demonstrate a method for determining the identity of enzyme-converted peptide products from single cells using a combination of capillary electrophoresis and liquid chromatography-mass spectrometry (LC-MS).

Key words: Single cell, Capillary electrophoresis, Chemical cytometry, Mass spectrometry, Enzyme activity, Fluorescent peptide

1. Introduction

Single-cell approaches for signal transduction network analysis are becoming more feasible as technology and methods advance, allowing more sensitive and specific detection of signaling events. Single-cell studies are particularly promising, since unlike bulk cell measurements which result in mixing of nonsynchronized heterogeneous responses, single-cell measurements portray accurately what is occurring in actual cells. As a result, in addition to their use for the analysis of rare or difficult to purify cells, single-cell studies are extremely useful for measuring complex, rapid or transient processes in cells. However, the detection and identification of the small quantities of analyte in single cells can pose a challenge due to the small quantities of total analyte. Although the detection of analyte can be simplified by the use of fluorescent labeling enabling

Sara Lindström and Helene Andersson-Svahn (eds.), *Single-Cell Analysis: Methods and Protocols*,
Methods in Molecular Biology, vol. 853, DOI 10.1007/978-1-61779-567-1_3, © Springer Science+Business Media, LLC 2012

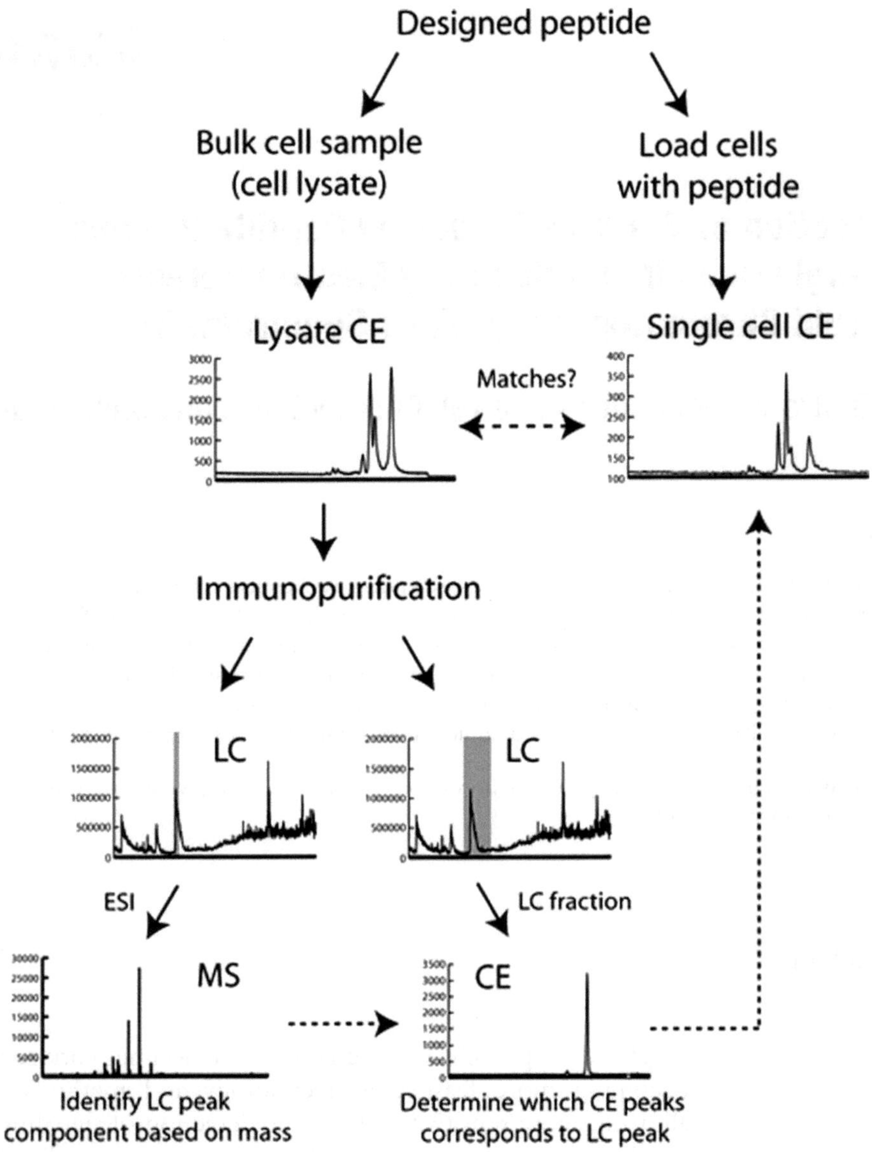

Fig. 1. Experimental workflow depicting the identification of enzyme processed peptides from single cells. First, single cells loaded with a designed fluorescently labeled peptide are analyzed using CE-LIF. Enzymatically converted peptide is then generated in larger amounts in cell lysate and analyzed by CE. Once conditions are found to yield similar cleavage product profiles as obtained in single cells, the peptide fragments are immunopurified and the masses of the peptide products in the lysate-based preparation are determined using LC-MS. The LC fractions are analyzed by both CE and MS, making it possible to assign a particular CE peak to a mass spectrum and thus identify the unknown peptide products from single cells. *Solid arrows* depict directional experiment workflow, and *dashed arrows* depict directional information flow. Figure reproduced from ref. 11 with permission from publisher.

up to single molecule detection, identification based on fluorescence measurements or migration time in electrophoretic separations can only yield clues to the identity of analyte. The use of fluorescent peptides to perform multiplexed signal transduction measurements for example has been employed by several groups for enzyme analysis (1–5); however, the identification of the peptide modifications which occur inside cells has been challenging. Identity has been determined by comigration of analyte with synthetic versions of the expected products, often in combination with modulation of the activity of the specific enzymes to be measured. In cases where comigration was not observed, all that could be determined is that the products were not those that were expected. Mass spectrometry (MS) measurements can yield high resolution determination of the molecular weight of enzymatic products providing far greater certainty in identification. However, use of mass spectrometers for monitoring single-cell enzymatic conversion remains extremely challenging due to the large number of background moieties in the cell as well as the low total amount of analyte due to the small size of samples. A method is presented here which uses capillary electrophoresis employing laser induced fluorescence detection (CE-LIF) for sensitive detection of enzyme modified fluorescent peptides, and then identifies these analytes by creating larger quantities of the same peptide products and identifying them using liquid chromatography (LC)-MS (experimental workflow diagrammed in Fig. 1).

2. Materials

All water was milli-Q filtered deionized water.

2.1. Fluorescent Peptide Cell-Loading

1. Cell media: Roswell Park Memorial Institute medium (RPMI) 1640 with L-glutamine supplemented with 10% fetal bovine serum.
2. Fluorescent peptide: Acetyl-GGVVIATVK(5-carboxyfluorescein)rrr-amide (where r represents D-Arginine) synthesized by Anaspec (Fremont, CA), was reconstituted at 4 mM in sterile DMSO.
3. Physiologic buffer: Supplement Hank's balanced salt solution (HBSS) (–Mg –Ca) with 20 mM HEPES, 100 μM $CaCl_2$, and 100 μM $MgCl_2$ adjust to pH 7.4 with 1 M NaOH, filter with 0.2-μM membrane syringe filter.

2.2. Capillary Electrophoresis (CE)

1. Run buffer: Add spermine to physiologic buffer to yield a final spermine concentration of 25 mM (see Note 1).
2. Capillary: A 30 μM inner diameter fused silica capillary (TSP 030375, Molex, Phoenix, AZ) cut to 40 cm in length with a ceramic scribe to produce a flat surface perpendicular to the length of the capillary, is inserted and glued into a bent 16 gauge blunt end needle (Stem cell technologies, Vancouver, BC) leaving around 1 cm of capillary exposed at the tip. For LIF detection a small window is burned in the polyimide coating using a lighter and the window is cleared of burnt polyimide

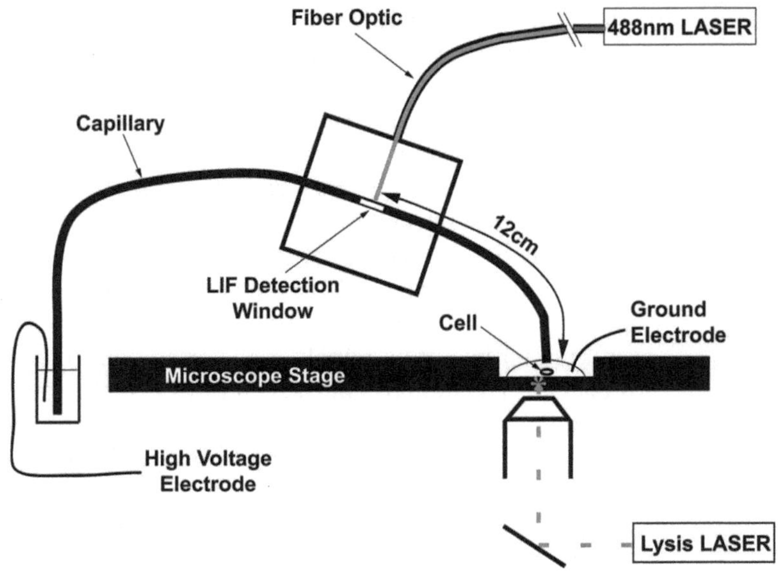

Fig. 2. Schematic of the microscope stage-mounted CE setup and single-cell lysis system. Figure reproduced from ref. 6 with permission from publisher.

using a methanol soaked Kimwipe. Initial conditioning of the capillary consists of filling the capillary with 0.1 M NaOH overnight, followed by a thorough rinse with distilled water.

3. Single-cell capillary electrophoresis instrument: The homebuilt instrument consists of a laser induced fluorescence detection system, as described (6, 7), mounted via a 3-axis translation stage (available from Thorlabs, Newton, New Jersey) to an inverted microscope (IX71, Olympus) (Fig. 2). The capillary is mounted onto the LIF detection system by feeding the capillary through the detection chamber and securing it in place via a Luer taper connection between the blunt end needle and a tapered projection on the LIF detection cube. The nonsampling end of the capillary is kept in a small buffer reservoir along with a high voltage platinum electrode mounted on a shuttle which can be raised or lowered along a rail for gravity loading of samples. The ground electrode is secured on the microscope stage. The electrodes are connected to a high voltage power supply (CZE 1000R, Spellman, Hauppage, NY) controlled by a computer terminal. For cell lysis, a nanosecond pulsed laser ($Nd:YVO_4$, Claire Lasers, Kitchener, ON) is directed through the side port of the microscope and redirected using a dichroic mirror through the objective lens (see Note 2).

4. TF1 cells: TF1 were expanded in cell media supplemented with 5 ng/mL human granulocyte-macrophage colony-stimulating factor (R&D Systems, Minneapolis, MN) in 25 cm^2 T-flasks in a humidified environment at 37°C with 5% CO_2.

2.3. Liquid Chromatography-Mass Spectrometry (LC-MS)

1. Pressure vessel with lid carrying a swagelock-fitting with teflon ferrule. The vessel is a home-built device of ~8 cm thick stainless steel block with a ~2 cm core for placing a 2-mL tube. The side of the block has an opening leading to a bleeder-valve connected to a low-grade helium cylinder using a high-pressure gas hose. The vessel can be closed from the top with a lid through six stainless steel threaded bolts. The bolts are tightened down evenly through six holes in the lid. The holes are typically manufactured in such way that they have a wide and narrow side and the lid has to be turned once the bolts are guided through the holes to tighten the lid safely. Also in the center of the lid a 1/8″ swagelock-fitting is added carrying a Teflon ferrule for holding the pulled tip during column packing process (see Note 3).
2. Buffer A: 95% water, 5% acetonitrile, 0.1% formic acid.
3. Buffer B: 95% acetonitrile, 5% water, 0.1% formic acid.
4. P-2000 laser-based micropipette puller (Sutter Instruments).
5. Proxeon nLC-EASY nanoflow HPLC system including cooled autosampler.
6. LTQ ion trap mass spectrometer (ThermoFisher Scientific, San Jose, CA) with nanoelectrospray ion source (Proxeon/ThermoFisher Scientific, Odense, Denmark).

3. Methods

3.1. Loading of Fluorescently Labeled Peptides into Cells

All centrifuge spins done at 800 × *g* for 3 min at 4°C. Note: Method is given for a specific peptide enzyme substrate; however, it should be widely applicable to other enzyme substrates with some optimization in run buffer and electrophoretic conditions.

1. Harvest TF1 cells at 0.5–1 × 10^6 cells/mL.
2. Spin down suspended cells and resuspend in Hanks buffered saline solution supplemented with 2% FBS at 10^5 cells per 100 μL on ice.
3. Add fluorescent peptide to concentration of 25 μM. Mix thoroughly but gently with pipette.
4. Add 100 μL of cell suspension to a precooled 2 mm gap electroporation cuvette (E-shot, Invitrogen, Carlsbad, CA).
5. Electroporate with a single 180V, 50 μF maximum capacitance pulse and return cuvette to ice bucket (see Note 4).
6. Add 300 μL of 37°C media to each cuvette and incubate for 5 min at room temperature.
7. Transfer cells into a prechilled low-adsorption polypropylene tube using a sterile transfer pipette and keep on ice.

8. Spin down and resuspend in 200 μL ice-cold physiologic buffer.
9. Repeat step 8 three more times, replacing the polypropylene tubes after the second wash (see Note 5).
10. Use cells immediately, less than 4 h if possible to avoid excessive peptide degradation due to protein turnover machinery in the cell and peptide efflux from the cell (see Note 6).

3.2. CE Analysis of Single Cells Loaded with Fluorescent Peptide

1. Before each run condition capillary with 0.1 M NaOH for 5 min, water for 1 min, 0.1 M HCl for 2 min followed by run buffer for at least 2 min.
2. Load 200 μL of room temperature physiological buffer onto the center of a number 1 coverslip held in place over the microscope objective (see Note 7).
3. Load 3 μL of the cell suspension (mix gently prior to taking cells) into the center of the physiological buffer on the coverslip.
4. Locate a cell which is sufficiently far from neighboring cells (>175 μM) that the capillary can be lowered around the cell without disturbing neighboring cells and position the cell in the center of the field of view.
5. Lower the capillary such that the targeted cell is contained within the lumen and the edge of the capillary is just touching the slide surface. You should be able to see the lumen clearly, and upon further lowering the cell will go out of focus.
6. Aspirate the buffer surrounding the cell and replace with 200 μL run buffer. Ensure that the ground electrode is immersed in run buffer.
7. Focus the microscope objective at the glass–buffer interface of the slide directly below the target cell.
8. Trigger the nanosecond laser at a pulse energy of ~3 μJ (see Note 8) and high voltage power supply set to supply a gradient of 230 V/cm (see Note 9).

3.3. Enzymatic Conversion of Fluorescent Peptide in Bulk Cell Lysate

1. Harvest TF1 cells at $0.5–1 \times 10^6$ cells/mL and resuspend in 14×10^6 cells/mL in ECB + 0.1% *n*-Dodecyl beta d-maltoside.
2. Add fluorescent peptide to a final concentration of 25 μM on ice and mix well.
3. Incubate reaction at 37°C and test aliquots at regular intervals using CE to obtain a significant portion of peptide conversion (>30%). Additional enzyme conversion may occur during the IP protocol (see Note 10).
4. Perform immunopurification of fluorescent peptide from the bulk cell lysate using Seize Primary Immunoprecipitation Kit (45335, Pierce, Rockford, IL) according to detailed instructions

provided with kit using anti-fluorescein/Oregon green antibody (A-889, Invitrogen, Carlsbad, CA).

5. Test immunopurified peptide sample to ensure the same peaks are all present in significant amounts by CE analysis.

3.4. Prepare Nanoflow Analytical Column

1. Cut a piece 40–60 cm of polyimide coated fused silica tubing with 100 μm ID and 365 μm OD (Molex, Phoenix, AZ) using a ceramic scribe.
2. Hold the center of the cut piece of fused silica tubing over an alcoholic laboratory burner and twist it while burning polyimide coating away of about 2–3 cm length.
3. Wipe pieces of burnt polyimide coating off with a methanol soaked Kimwipe.
4. Mount fused silica into laser puller with coating free area in the shielded center area of focused laser beam.
5. Use predefined column pulling program to pull tip (Table 1) (see Note 11).
6. Add about 1 mg of column packing material (silica based C18 coated beads with diameter of 3 μm) to 500 μL HPLC-grade methanol to make a slurry in a round bottom 2-mL sample tube.
7. Add a 3-mm stir bar to the sample tube and place in a specialized pressure vessel on a magnetic stirrer.
8. Turn on the magnetic stirrer and ensure mixing of slurry before closing the pressure vessel securely with the lid.
9. Rinse the pulled 100 μm ID column tip with methanol before placing it blunt end first through the opening in the pressure-vessel lid. Leave a gap of around 3 mm (by feel) to the bottom of the tube.

Table 1
Four-step pulling program for a 100 μm i.d. × 365 μm o.d. fused silica tubing on a Sutter Instruments Co. laser puller model P-2000

HEAT	FIL	VEL	DEL	PUL
280	0	30	200	0
280	0	30	200	0
260	0	30	200	0
250	0	30	200	0

10. Secure the column by tightening the nut pressing into the Teflon ferrule so that the column will not move when pressure is applied (see Note 12).
11. Apply 600–1,000 psi pressure to the securely tightened pressure vessel with using low grade helium.
12. Leave column packing until the bed reaches a length of 15 cm (see Note 13).
13. When column packing is complete, release the pressure from vessel through a bleeder-valve and remove the column.
14. Mount packed column on a nanoLC-system and apply isocratic flow with 0.5–1 μL/min of a mixture 85% buffer B and 15% buffer A. Let it pressurize for about 10–20 min. Make sure the pressure is larger than 150 bar.
15. Apply isocratic flow as in step 14 for a mixture of 5% buffer B and 95% buffer A for about 10–20 min using a flow rate of 0.5–1.0 μL/min. Again pressure should be higher than 150 bar.
16. When conditioning is finished, cut the column 1–2 cm behind the packing material and mount it in the nanoLC-flow line on a nanoESI source (Proxeon/ThermoFisher).

3.5. Perform nanoLC-MS and Interpret LC-MS Data Using a Linear Ion Trap

1. The following instructions are valid when working on a LTQ ion trap (ThermoFisher Scientific, San Jose, CA).
2. The column setup was done without using a precolumn. Analytical column (i.e., pulled tip with packing material) was directly connected on its front-end to the nLC-EASY outline using a PEEK-Tee junction (Upchurch Scientific), containing the high-voltage connection. The back-end of the column (pulled tip) was positioned 3 mm distal from the LTQ-orifice.
3. Instrument Parameters:

   (a) Proxeon nLC-EASY program:

   Sample pickup: 1 μL

   Sample loading: 2 μL

   Autosampler needle wash: 100 μL 100% water, 0.1% formic acid (see Note 14).

   Solvent gradient: for 2 min hold at 2% buffer B, 22 min gradient from 2 to 24% buffer B, 6 min from 24 to 90% buffer B, then hold for 5 min at 90% buffer B, gradient for 1 min from 90 to 2% buffer B, and finally hold at for 8 min at 2% buffer B.

   Flow rate: 500 nL/min throughout the whole gradient program.

   (b) Ion trap parameters:

   In instrument parameter section of Xcalibur v2.0.6 define a precursor scan from *m/z* 400 to 2,000 and set ion trap target values at 3×10^4 (see Notes 15 and 16).

4. Define the sample pickup-position in a 96-well microtiter plate from Proxeon nLC-EASY autosampler. Start the corresponding ion trap instrument method by a contact closure sent from the nLC-EASY instrument at "Start Gradient" position, defined in the configuration section of Proxeon software accessed via touch-screen interface (see Note 17).
5. Upon completion of the run, examine the chromatogram for major peaks. Devise a fraction collection timing scheme to collect each major peak in a separate fraction. Chromatogram sections which do not contain peaks can be broken into 5-min long fraction collection periods (see Note 18).

3.6. Sample Fractionation from Nanoflow Column for Further Analysis with CE

1. Apply gradient as used above with the difference that end of the column (tip) is not mounted upfront the mass spectrometer, but pointing into a small prelabeled low-adsorption centrifuge tube or microwell container. No high voltage is applied!
2. Collect fractions from column tip by gently touching the forming droplet to the wall of the tube/microwell at the end of the fraction collection period (see Note 19).
3. Directly after each fraction collection, gently drag a dry Kimwipe along the tip in the direction of the point to remove any remaining liquid from the previous fraction.
4. Fractions under 2 μL in volume should be diluted with run buffer to avoid sample drying, and to ensure that sufficient volume is present to allow sample transfer to the CE system.

3.7. CE Analysis of Collected LC Fractions

1. Before each run condition flush capillary with 0.1 M NaOH for 5 min, water for 1 min, 0.1 M HCl for 2 min followed by run buffer for at least 2 min.
2. Transfer 0.5–1 μL of sample to flat surface of a clean low-adsorption microtube cap on the microscope stage (see Note 20).
3. Lower the capillary inlet tip into the sample and lower the outlet reservoir 10 cm for 10 s.
4. Transfer the capillary tip into a cap filled with run buffer and begin CE run with the high voltage power supply delivering a voltage gradient of 230 V/cm.
5. Compare migration times of peaks obtained from each fraction with the migration time of peaks obtained from single cells and lysate to correlate LC peaks with CE peaks. Once determined, the identification of the components in the LC peaks can be determined through analysis of the mass spectra obtained in those peaks during the LC-MS run (see Fig. 3).

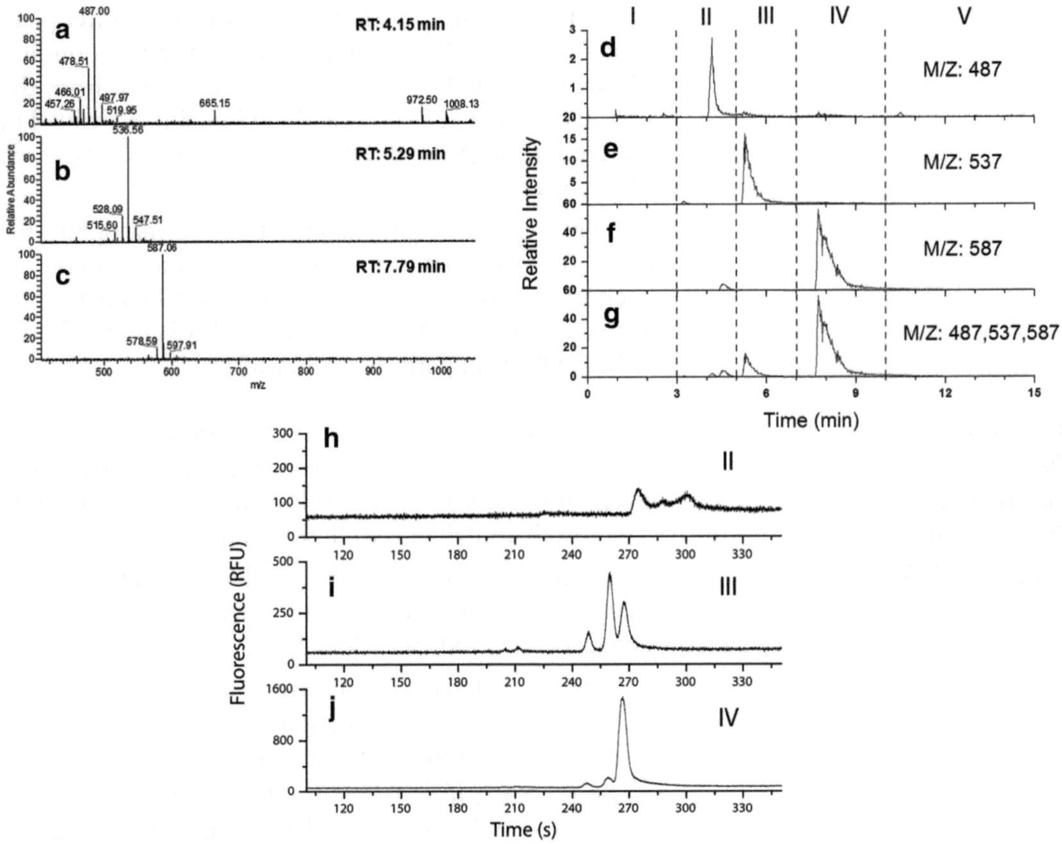

Fig. 3. Mass spectra at the chromatographic apex of individual major peaks (**a**–**c**), RT indicates the retention time corresponding to the specific mass spectrum. LC-MS extracted ion chromatograms and CE analysis of LC fractions of the antifluorescein immunopurified eluate. Extracted ion chromatograms of the mass to charge ratios (*m/z*) 487, 537, and 587 representing the calculated *m/z* for peptide cleavage products of 3, 4, and 5 amino acid long, alone (**d**–**f**, respectively) and together (**g**) from the LC-MS run. LC fraction collection periods for CE analysis are denoted by roman numerals *I*–*V* separated by *vertical dashed lines* (**d**–**g**). Capillary electropherograms of LC fraction II (**h**); LC fraction III (**i**); and LC fraction IV (**j**). Portions of figure reproduced from ref. 11 with permission from publisher.

4. Notes

1. Make fresh daily; after a few days a precipitate will become visible.
2. Use of a pulse laser to lyse the cell of interest is used; however, other methods can be substituted as described in ref. 8. An alternate LIF detection system design for use with capillary electrophoresis is given in (9).
3. Pressure vessels can be purchased from commercial vendors, e.g., Proxeon/ThermoFisher Scientific, or plans to build home-made device can be obtained from several laboratories using this technology (10).

4. The voltage used will depend on the cell type based on membrane composition and cell size.
5. Following the washes ensure that there is no significant levels of peptide in the buffer surrounding cells by running a sample of cell supernatant without cells by CE.
6. An aliquot of cells may be used to determine level of peptide loading based on flow cytometry measurements before CE to ensure loading is sufficient.
7. Physiological buffer should be at room temperature when applied to the coverslip to avoid condensation on the bottom of the glass.
8. The optimal laser pulse energy will depend on the specific laser used and the optical setup (6). To determine the optimal pulse energy for a specific instrument fire a pulse at the coverslip–buffer interface and modulate the pulse energy to find the minimum pulse energy which consistently yields divots in the coverslip.
9. The optimal voltage depends on buffer/capillary system used, ideally a joule curve should be performed to determine the maximum voltage that will not lead to significant joule heating.
10. Ensure that peaks observed represent the peaks observed from single cell CE measurements. Cells may be loaded in a similar fashion as performed for single cell analysis and aggregated before lysis if sufficient peptide is available to load cells in this manner.
11. Suggested column pulling program, settings might vary slightly on each machine and have to be individually optimized.
12. Wear safety glasses and do not bend over the pressure vessel when applying pressure.
13. Packing can be monitored best against a spotlight shining from behind the column towards the operator. If pulled tip is not open, a flat side of the scribe can be used to gently touch the pressurized tip from the side, which in most cases results in opening the tip.
14. Buffer A may be used instead of formic acid for needle wash.
15. SIM-scans (or zoom-scans), due to their higher resolution, may be performed for five mass unit ranges encompassing expected peptide products or after an initial LC-MS run as validation to define the charge state of detected peaks and limit false identifications.
16. No precolumn/analytical column equilibration has been applied.
17. Alternatively to set up the gradient and start the run, a Proxeon nLC-EASY driver can be obtained as Xcalibur plugin.
18. If peaks are packed too densely on the chromatogram, the LC-MS run should be repeated with modifications in the solvent gradient or, if necessary, with different separation media.

19. Column tip is fragile, so try not to touch the actual tip with the tube/microwell to avoid breakage.
20. Do not touch the top surface of the cap if possible, any dirt/oil will cause the small droplet to spread on the cap and lead to adsorption of sample.

References

1. Arkhipov, S. N., Berezovski, M., Jitkova, J., and Krylov, S. N. (2005) Chemical cytometry for monitoring metabolism of a Ras-mimicking substrate in single cells, *Cytometry A 63*, 41–47.
2. Li, H., Sims, C. E., Kaluzova, M., Stanbridge, E. J., and Allbritton, N. L. (2004) A quantitative single-cell assay for protein kinase B reveals important insights into the biochemical behavior of an intracellular substrate peptide, *Biochemistry 43*, 1599–1608.
3. Meredith, G. D., Sims, C. E., Soughayer, J. S., and Allbritton, N. L. (2000) Measurement of kinase activation in single mammalian cells, *Nat Biotechnol 18*, 309–312.
4. Sims, C. E., Meredith, G. D., Krasieva, T. B., Berns, M. W., Tromberg, B. J., and Allbritton, N. L. (1998) Laser-micropipet combination for single-cell analysis, *Anal Chem 70*, 4570–4577.
5. Zarrine-Afsar, A., and Krylov, S. N. (2003) Use of capillary electrophoresis and endogenous fluorescent substrate to monitor intracellular activation of protein kinase A, *Anal Chem 75*, 3720–3724.
6. Brown, R. B., and Audet, J. (2007) Sampling efficiency of a single-cell capillary electrophoresis system, *Cytometry A 71*, 882–888.
7. Brown, R. B. (2010) Developing methods to enable multiplexed signal transduction measurements in single cells, PhD Thesis, p 144, University of Toronto, Toronto.
8. Brown, R. B., and Audet, J. (2008) Current techniques for single-cell lysis, *J R Soc Interface 5 Suppl 2*, S131–138.
9. Poe, B. G., 3rd, Duffy, C. F., Greminger, M. A., Nelson, B. J., and Arriaga, E. A. Detection of heteroplasmy in individual mitochondrial particles, *Anal Bioanal Chem 397*, 3397–3407.
10. Gatlin, C. L., Kleemann, G. R., Hays, L. G., Link, A. J., and Yates, J. R., 3rd. (1998) Protein identification at the low femtomole level from silver-stained gels using a new fritless electrospray interface for liquid chromatography-microspray and nanospray mass spectrometry, *Anal Biochem 263*, 93–101.
11. Brown, R. B., Hewel, J. A., Emili, A., and Audet, J. (2010) Single amino acid resolution of proteolytic fragments generated in individual cells, *Cytometry A 77*, 347–355.

Chapter 4

Laser Capture Microdissection for Gene Expression Analysis of Specific Cell Populations in Single Blastocysts*

Ward De Spiegelaere, Muriel Filliers, and Ann Van Soom

Abstract

Laser capture microdissection (LCM) allows for the isolation of small tissue fractions from heterogeneous tissue sections, for downstream genetic or proteomic analysis without contamination by the surrounding tissue. This technique can also be successfully used for the isolation of small tissue fractions from developing embryos, such as expanding blastocysts. However, the small size of early-stage embryos hampers tissue processing prior to LCM. The present protocol describes the application of LCM to isolate specific cell fractions from blastocysts for downstream gene expression analysis with RT-PCR.

Key words: Laser capture microdissection, Gene expression, Blastocysts, RT-PCR

1. Introduction

Comparing gene expression patterns of specific cell fractions in early embryos may contribute to our understanding of initial development and differentiation. As an example, the isolation of pure, homogeneous inner cell mass cells (ICM) and trophectoderm cells (TE) from a single embryo (blastocyst) is a crucial first step to obtain more information of their expression profile. As early as 1972, microsurgery has been used to mechanically dissect these distinct cell populations (1). Recently, laser-assisted dissection of ICM cells has been used as an effective strategy to cultivate embryonic stem cells (2). Unfortunately, these methods cannot prevent contamination of the ICM cell population with TE cells. As such, the specific gene expression pattern of ICM cells cannot be inferred with these techniques. Two methods have been described to

*Ward De Spiegelaere and Muriel Filliers contributed equally to this work.

Sara Lindström and Helene Andersson-Svahn (eds.), *Single-Cell Analysis: Methods and Protocols*, Methods in Molecular Biology, vol. 853, DOI 10.1007/978-1-61779-567-1_4, © Springer Science+Business Media, LLC 2012

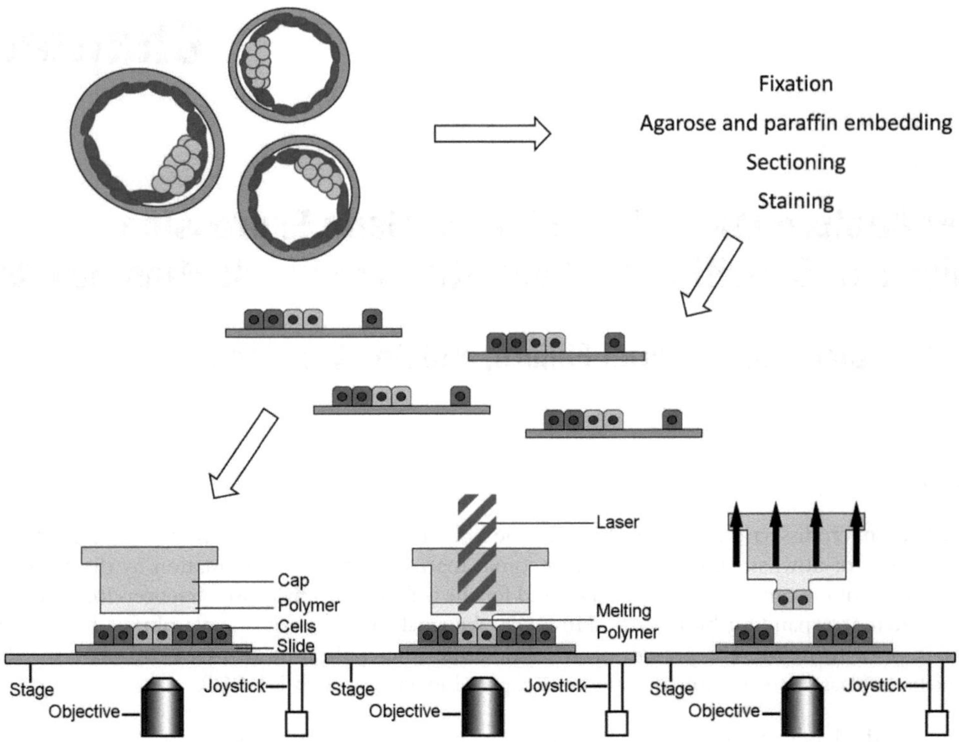

Fig. 1. Schematic illustration of the LCM procedure.

overcome this problem, i.e., immunosurgery and magnetic activated cell sorting. During immunosurgery the outer TE cells are selectively killed, releasing pure ICM cells (3). Immunosurgery has been used to depict comparative molecular portraits of ICM and TE cells in human blastocysts (4). The magnetic activated cell sorting technique uses antibodies that are conjugated to magnetic microbeads for subsequent cell sorting (5). However, during the processing steps of the latter two techniques, the living blastocysts are exposed to chemical substances. These substances may cause stress on the embryonic cells. Moreover, it has been observed that this can change the expression profile of the cells (6).

Laser capture microdissection (LCM) is a promising alternative to the previous techniques, as it provides the possibility to isolate pure samples of ICM cells, without contamination by the surrounding tissue, and without changing the gene expression pattern of the cells (7). Although LCM with frozen sections is optimal for preserving an adequate RNA quality, the small size of blastocysts impedes tissue handling and sectioning of these embryos. As a result, a protocol was optimized in which cultured bovine blastocysts are embedded in an agarose gel prior to paraffin embedding. This allows tissue processing for subsequent LCM (Fig. 1) and gene expression analysis with the reverse transcription polymerase chain reaction technique (RT-PCR) (8). In addition, the LCM technique might also be used for downstream next-generation sequencing (9).

2. Materials

One of the major obstacles with RT-PCR is the preservation of good quality RNA (10). These molecules are extremely susceptible to degradation, and are easily degraded by RNase enzymes that can be endogenously present in the sample, but can also be introduced in the pre-PCR processing steps. Consequently, to prevent contaminating the samples with RNases, all recipients, products, and solutions should be RNase free (see Note 1).

2.1. Tissue Fixation and Embedding

1. Wash buffer: phosphate buffered saline solution: add 0.2 g KH_2PO_4, 0.2 g KCl, 8 g NaCl, and 2.2 g $Na_2HPO_4 \cdot 7H_2O$ to 800 ml distilled water, adjust to pH 7.4 with 100 mM NaOH and adjust the volume to 1 l (see Note 2).
2. Fixative: Modified methacarn solution: Mix 10 ml acetic acid in 80 ml methanol. This solution can be stored at room temperature (see Note 3).
3. 2% Agarose: Mix 2 g of agarose with 100 ml distilled RNase-free water. Heat to 60°C and vortex to dissolve the agarose. This solution can be stored at room temperature.
4. Petri dish (∅ 35 mm).
5. Dissecting microscope.
6. 0.5-ml Microcentrifuge tubes.
7. 80, 96, and 100% ethanol; xylene; paraffin for tissue embedding.
8. Biopsy cassettes.
9. Metal base molds for paraffin embedding.

2.2. Sectioning and Staining

1. Rotary microtome for sectioning paraffin-embedded tissue samples.
2. Warm water bath with 0.7% gelatin solution in ultrapure RNase-free water at 49°C (see Note 4).
3. Microscopic glass slides 26 mm × 76 mm.
4. Hot plate.
5. 0.1% Cresyl violet in 85% ethanol solution.
6. 95 and 100% ethanol.
7. Xylene.

2.3. Laser Capture Microdissection

1. Pixcel® IIe LCM device (Applied Biosystems) or a similar LCM device.
2. Capsure® Macro LCM caps.
3. RNA isolation kit: Commercial RNA isolation kits are available to isolate total RNA from small samples.

4. RNA quality assessment: Experion™ Automated gel electrophoresis system, with RNA High sense chips, which are specially suited for small amounts of RNA. Similar devices working with lab-on-a-chip microfluidics to analyze total RNA-integrity are also available on the market.

2.4. Reverse Transcription Polymerase Chain Reaction

1. Reverse transcription kit: Commercial kits are available.
2. Control primers for bovine trophectoderm cells: KRT18 gene (Primers: 5′-GCAGACCGCTGAGATAGGA-3′ and 5′-GCAT ATCGGGCCTCCACTT-3′; 144 bp; annealing temperature 62°C) (8).
3. Positive control primers for bovine cells: 18 S rRNA (5′-AGAAACGGCTACCACATCCA-3′ and 5′-CACCAGAC TTGCCCTCCA-3′; 169 bp; annealing temperature 62°C) (8).
4. Gene-specific primers.
5. PCR master mix: Commercial kits are available.

3. Methods

The tissue processing steps before RNA isolation can be detrimental for RNA quality in the sample due to the action of RNase enzymes. As such, apply RNase inactivating solution (see Note 5) to all surfaces, wipe dry and rinse with RNase-free water to remove the decontamination solution before starting the protocol. Try to reduce processing times, especially when the samples are in an aqueous solution. RNases are active in aqueous solution, but in a dry state or in solutions with more than 70% alcohol, they are less active. In addition, working with gloves during the entire procedure will further prevent contamination with RNases.

3.1. Tissue Fixation and Processing

1. Collect the blastocysts and wash them three times in PBS (see Note 6).
2. Place the blastocysts in a small petri dish (∅ 35 mm), which is filled with the modified methacarn fixative. Fix the blastocysts for 24 h at 4°C.
3. Transfer the blastocysts in a 0.5-ml microcentrifuge tube with a minimal amount of fixative, using a dissection microscope.
4. Heat the 2% agarose until it almost starts boiling (see Note 7) and add approximately 50–100 µl hot agarose to the microcentrifuge tube (see Note 8). Carefully pipette this solution up and down to mix the agarose solution with the fixative and the blastocysts.
5. Leave the centrifuge tubes for 20 min at 4°C, to let the agarose gel polymerize.

6. Remove the agarose gel from the microcentrifuge tube. This can best be performed by carefully cutting off the tip of the microcentrifuge tube. Consequently, the agarose gel can be pushed out of the microcentrifuge tube by carefully pushing the gel with the tip of a blunt forceps through the opening of the tube.
7. Place the agarose gel in a biopsy cassette and label the cassette with a marker that is resistant to alcohol and xylene (see Note 9).
8. Tissue processing: Place the biopsy cassettes consecutively in 80% ethanol for 1 h, 96% ethanol (two baths, 1 h each), 100% ethanol (two baths, 1 h each), xylene (two baths, 1 h each), and finally in 4 baths of paraffin at 60°C for 40 min each (see Note 10).
9. Embed the tissue by placing the agarose gel in a metal base mold, fill the mold with paraffin, and mount the biopsy cassette on top. The latter can be used to fix the paraffin block to the microtome.

3.2. Tissue Sectioning and Staining

1. Make serial sections of the agarose gels at 10 μm per section (see Notes 11 and 12). Let the sections float on the hot gelatin solution for 30 s, to stretch them. Subsequently, place the sections on the glass slides.
2. Leave the glass slides on a hot plate (60°C) for 5 min to adhere the sections to the glass slides (see Note 13).
3. Deparaffinize the sections by placing them in two successive baths of xylene for 3 min. Subsequently, dip the sections shortly in a 100 and 95% ethanol bath.
4. Stain the sections with the cresyl violet solution for 30 s (see Note 14).
5. Dehydrate the sections by shortly dipping them in a 95 and 100% ethanol bath and subsequently in two successive baths of xylene for 3 min. Subsequently, place the sections in the fume hood to let the xylene evaporate (approximately 10 min) (see Note 15).

3.3. Laser Capture Microdissection and RNA Isolation

1. Optimize the laser spot size by lowering the laser power and the pulse width (see Note 16).
2. Fire the laser on the cells that should be isolated, and subsequently lift the cap to isolate them (Fig. 2). In order to get more sample material, the same cap can be used in subsequent sections of a serially sectioned blastocyst (see Note 17).
3. Place the cap on a 0.5-ml microcentrifuge tube (see Note 18) containing the lysis buffer from the RNA isolation kit, and place the tube upside down at the conditions recommended by the manufacturer of the RNA isolation kit.
4. Isolate the total RNA from the sample with the RNA isolation kit.

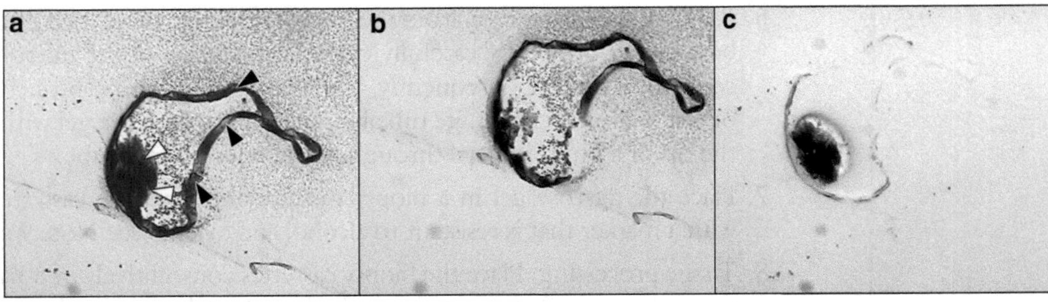

Fig. 2. Sections of an expanded bovine blastocyst stained with cresyl-violet before (**a**) and after microdissection of the ICM cells (**b**). The ICM cells (*white arrowheads*) are surrounded by TE cells (*black arrowheads*; (**a**)). After microdissection, the isolated cell fraction on the cap can be visually controlled to ascertain the absence of contamination by TE cells (**c**).

5. Control the quality of RNA with the Experion™ Automated Electrophoresis System and proceed to the reverse transcription. However, in some cases, the amount of RNA will be too small to check the quality with "lab on a chip" automatic electrophoresis systems. In that case, the 3′-5′ ratio assay can be used to deduce the quality of the RNA as described by Nolan et al. (11). An additional advantage of the 3′-5′ ratio is that it can also be used after amplification of the RNA or cDNA.

3.4. Reverse Transcription and Polymerase Chain Reaction

1. Reverse transcribe the RNA to single stranded coding DNA (cDNA).
2. The cDNA can be used in the polymerase chain reaction (PCR). When using bovine blastocysts, the KRT18 primer pair serves as an internal control in inner cell mass cells to detect possible contamination from trophectoderm cells, as the KRT18 gene is only expressed in TE and not in the cells of the ICM (8). The S18 rRNA primer pair (or a primer pair for another suitable reference gene) is used as a positive control, to ensure that the isolation and pre-PCR processing steps worked well in all samples. However, for quantitative PCR (RT-qPCR) normalization should be performed with more than one reference gene, to prevent bias due to biological variation in expression profiles of the reference genes (12, 13).

4. Notes

1. Glass recipients can be dry-heat-sterilized in an oven at 180°C for a minimum of two hours; RNase-free water can either be purchased or self-made by running ultrapure water through a nuclease filter which retains RNases, or by DEPC treatment of distilled water. All recipients and devices that cannot be sterilized can be wiped with an RNase inactivating solution.

2. Instead of PBS, another physiological salt solution can also be used to wash the blastocysts.
3. The modified methacarn solution provides good morphological quality of the samples while maintaining an adequate RNA quality for downstream RT-PCR. However, other fixatives may also be used, see Buesa et al. (14).
4. A 0.7% gelatin solution is used to adhere the section to the glass slides. Other methods are also available to adhere the sections to the slides, such as using pre-coated slides. However, care should be taken that the adhesion is not too strong, as this could prevent the isolation of the sample material in the subsequent LCM.
5. Multiple RNase activating solutions are commercially available, but a 100 mM NaOH solution in EDTA works as well.
6. Occasionally, some blastocysts can be lost during processing. Placing several blastocyst together (for example: 15) will ensure that each sample contains blastocysts, and will ease the downstream processing.
7. It is important that the agarose gel is hot enough so that it is still above the melting point after mixing with the fixative. If the agarose is too cold, it will polymerize too quickly and the blastocysts might not be totally embedded in the agarose gel. As such, some blastocysts may be lost during the subsequent processing steps.
8. The amount of agarose is optional. We generally try to minimize the amount of agarose, to have a high density of embedded blastocysts, easing the subsequent tissue processing. However, the amount of agarose to fixative should at least be 50/50. A lower ratio might impede a proper polymerization of the agarose gel.
9. Do not label the biopsy cassette with an alcohol based permanent marker, as the ink will dissolve in the alcohol baths during tissue dehydration. When no adequate marker is available, a small paper labeled by pencil can be inserted in the biopsy cassette, together with the agarose gel.
10. Tissue processing can be executed manually, but automatic tissue processing devices are also available. 100% isopropyl alcohol can be used instead of 100% ethanol. In addition, xylene may also be replaced by toluene.
11. The thickness of the sections is optional. More material can be isolated when using thick sections, but may compromise the visual quality of the sections.
12. Placing more sections on one slide will ensure a faster staining procedure and will ease the downstream LCM procedure. Up to ten sections can normally be placed on one glass slide.

13. Check whether the paraffin surrounding the tissue is melting. This is necessary for a good adhesion on the slide; if the paraffin does not melt, increase the temperature of the hot plate or leave the samples on the hot plate for a longer time.
14. Instead of cresyl violet, other stains can be used as well, including eosin or methyl green. Haematoxylin staining can also be used, but this stain only works well in an aqueous solution and might enable endogenous RNases to degrade the RNA.
15. Do not use the same baths of xylene for the deparaffination and the dehydration steps. In addition, use fresh solutions every time. Especially the solutions for the dehydration should be pure, as an incomplete dehydration may cause a too tight adherence of the samples to the glass slides, impeding the sample isolation with LCM.
16. The following settings worked well in previous assays: size lever at 7.5 μm, threshold voltage at 190 mV, laser pulse power at 70 mW and pulse width at 0.5 ms. However, these settings may depend on the type of LCM device. As such, these settings should be optimized and tested before every isolation. In order to avoid excessive use of caps, these settings can be tested on an empty space under the cap.
17. Occasionally, some neighboring material may stick to the cap. This signifies that the sections do not adhere well enough to the glass slide. However, when a valuable sample is at stake, the extra material can be removed by carefully pushing the sticky side of a Post It® note on the underside of the cap. This will remove the extra material without affecting the cells on the region of interest. Be sure to check the cap under the microscope for a second time to ensure that the region of interest is still on the cap.
18. Before starting, check whether the cap fits well on the microcentrifuge tubes because, depending on the supplier, some tubes have an opening that is too wide or too narrow.

Acknowledgments

The authors thank Jurgen De Craene, Isabel Lemahieu, and Petra Van Damme for their excellent technical assistance and Paul Simoens for his constructive comments and critical review of the manuscript. M. Filliers is a research fellow of the Fund for Scientific Research–Flanders, Belgium (FWO), aspirant 1.1.477.07N00.

References

1. Gardner R L (1972) An investigation of inner cell mass and trophoblast tissues following their isolation from the mouse blastocyst. J Embryol Exp Morphol **28**:279–312
2. Tanaka N, Takeuchi T, Neri Q V et al (2006) Laser-assisted blastocyst dissection and subsequent cultivation of embryonic stem cells in a serum/cell free culture system: application and preliminary results in a murine model. J Trans Med **4**:20
3. Solter D and Knowles B B (1975) Immunosergery of mouse blastocysts. Proc Natl Acad Sci **72**:5099–5102
4. Adjaye J, Huntriss R, Herwig R et al (2005) Primary differentiation in the human blastocyst: Comparative molecular portraits of inner cell mass and trophectoderm cells. Stem Cells **23**:1514–1525
5. Ozawa M and Hansen P J (2011) A novel method for purification of inner cell mass and trophectoderm cells from blastocysts using magnetic activated cell sorting. Fertil Steril **95**:799–802
6. Cauffman G, Van de Velde H, Liebaers I et al (2005) DAZL expression in human oocytes, preimplantation embryos, and embryonic stem cells. Mol Hum Reprod **11**:405–411
7. De Spiegelaere W, Cornillie P, Van Poucke M et al (2011) Quantitative mRNA expression analysis in kidney glomeruli using microdissection techniques. Histol Histopathol **26**: 267–275
8. Filliers M, De Spiegelaere W, Peelman L et al (2011) Laser capture microdissection for gene expression analysis of inner cell mass and trophectoderm from blastocysts. Anal Biochem **408**:169–171
9. Takahashi H, Kamakura H, Sato Y et al. (2010) A method for obtaining high quality RNA from paraffin sections of plant tissues by laser microdissection. J Plant Res **123**:807–813
10. Fleige S and Pfaffl M (2006) RNA integrity and the effect on the real time qRT-PCR performance. Mol Aspects Med **27**:126–139
11. Nolan T, Hands R E and Bustin S A (2006) Quantification of mRNA using real-time RT-PCR. Nat Protoc **1**:1559–1582
12. Goossens K, Van Poucke M, Van Soom A et al (2005) Selection of reference genes for quantitative real-time PCR in bovine preimplantation embryos. BMC Dev Biol **5**:27
13. Vandesompele J, De Preter K, Pattyn F et al (2002) Accurate normalization of real-time quantitative RT-PCR data by geometric averaging of multiple internal control genes. Genome Biol **3**:RESEARCH0034
14. Buesa R J (2008) Histology without formalin? Ann Diagn Pathol **12**:387–396

Part II

Miniaturized Solutions

Chapter 5

Single-Cell Culture in Microwells

Sara Lindström and Helene Andersson-Svahn

Abstract

In order to better understand cellular processes and behavior, a controlled way of studying high numbers of single cells and their clone formation is greatly needed. This chapter describes a microwell plate with 672 wells in a standard array/slide format, applied for single-cell culture and analysis. Single cells can be seeded into each well of the plate (1) manually or (2) automatically using a sorting flow cytometer, followed by week-long culture and detection of cell growth, protein expression, etc. The glass/silicon plate is compatible with most standard instrumentation to facilitate easy handling and enable use of the plate for single-cell analysis in most laboratory settings.

Key words: Single cell, Microwell, Array, Plate, FACS, Long-term analysis, Clone formation, Cell culture

1. Introduction

When studying cells, one by one, a true detailed picture of single-cell behavior is given. It can for example be interesting to monitor a cell as it migrates or divides into two cells, how many divisions that occur over time, and the rate of cell division in a specific cell clone. Cell divisions are arousing attention in for example cancer, due to the uncontrolled rate of cell proliferation of a cancer cell as compared to a normal cell. The dynamic study of living cells can increase the understanding of the interconnecting molecular events continually taking place in each cell. Each cell is more or less different from the other, even within the same cell type (1). Cellular heterogeneity is well known in bacteria and increasingly apparent in eukaryotic cells (2). Previously, cell cultures have been considered to be quite homogeneous in their nature, and the analysis of a collection of cells has been believed to give an accurate assessment of the behavior of the cells in that culture or tissue. The average response of the cells was, and often still is, interpreted as the

Sara Lindström and Helene Andersson-Svahn (eds.), *Single-Cell Analysis: Methods and Protocols*, Methods in Molecular Biology, vol. 853, DOI 10.1007/978-1-61779-567-1_5, © Springer Science+Business Media, LLC 2012

response of all cells in that sample. Additionally, the effect of neighboring cells (signaling, interactions etc.) on cell behavior is often neglected due to difficulties in monitoring such detailed phenomena. The attitude to ensemble measurements is starting to change and today's researchers are increasingly aware of, hence interested to study, the resulting effects of heterogeneity in cell samples, as mirrored in recent reviews on single-cell analysis (3–8).

One way of isolating individual cells is to mechanically separate the cells by physical boundaries, i.e., wall structures. Wells on a plate enable parallel analysis of multiple samples and have long been applied in life science. Depending on the application, different modifications have been performed. For example, standard 96-well plates with different bottom materials have been developed for improved transparency and reduced background flourescence, when detecting cell growth using a microscope. By miniaturizing, more samples (wells) can be run in parallel leading to a higher throughput. Microwells can in many cases be considered a simple and straightforward technique, in terms of theory, practical handling and fabrication. Depending on the intended application, a microwell device can be designed in numerous different ways by selecting different parameters such as shape, size, number of wells, etc. as exemplified in Table 1.

The method presented here is long-term culture of single cells as they expand into clones. A microwell plate consisting of silicon walls, a coverslip glass bottom, and a breathable culture top membrane is used for the experiments. The plate holds 672 microwells and single cells are sorted into the 0.5 μl-wells using a flow cytometer. Microscopic detection is used throughout the culturing. The method is very general and can be applied to many different types of cells. Combined with a wide range of end-point detection methods (fluorescence, bright field, magnetic beads, etc.) and the possibility to run many different assays on the investigated cells (antibody labeling, PCR, microfluidics, etc.), it should be quite easy to adapt and adjust the method, dependent on the requested study.

2. Materials

2.1. Microwell Plate

1. The microwell plate and the cell culture membrane are available at http://www.picovitro.com.
2. For fabrication of the microwell plate and cell culture membrane, use standard microfabrication protocols in a dedicated clean-room environment; more details on the design and fabrication can be found elsewhere (9).
3. For the entire protocol, three microwell plates and 2 cell culture membranes are needed.

Table 1
Overview of a number of parameters in microwell chip designs, summarized from references: (11–16)

	Shape of wells			Material of wells						Size of wells			Number of wells		
References	Square	Hexagonal	Round	Glass	Silicon	PDMS	Fiber	SU-8	PEG	<30 μm	ca. 50 μm	>100 μm	100s	10,000s	100,000s
Taylor and Walt (11)		✓					✓			✓			✓		
Chin et al. (12)	✓							✓				✓		✓	
Revzin et al. (13)	✓								✓	✓				✓	
Rettig and Folch (14)			✓			✓				✓	✓			✓	
Deutsch et al. (15)		✓		✓						✓				✓	
Tokimitsu et al. (16)			✓		✓					✓					✓

4. Steam sterilizable plastic bags (Wipak Medical Industrial Packaging, Steriking R41).
5. Autoclave the microplate for sterile cell culture environment.

2.2. Cell Culturing and Staining

All phosphate-buffered saline (PBS) was filtered and sterilized (autoclaved).

1. Nonadherent cells: Leukemic K-562 cells (European Collection of Cell Cultures, ECACC No. 89121407).
2. RPMI 1640 media with 2 mM L-glutamine, supplemented with 10% fetal bovine serum (FBS) and 1% antibiotic/antimycotic solution (all from Invitrogen).
3. Adherent cells: Epithelial U-2 OS cells (ECACC No. 92022711).
4. McCoy's 5a medium (modified) with 1.5 mM L-glutamine supplemented with 10% FBS and 1% antibiotic/antimycotic solution (all from Invitrogen).
5. Cells were maintained in standard Petri dishes (100×20 mm^2, BD Falcon).
6. 25 μg/ml fibronectin (Sigma).
7. 0.05% trypsin/0.53 mM EDTA (Invitrogen).
8. PBS buffer: 10 mM Na_2HPO_4, 137 mM NaCl, 1.8 mM KH_2PO_4, 2.7 mM KCl, pH 7.5. For Fluorescence-Activated Cell Sorting (FACS) experiments, the PBS buffer was diluted four times (referred to as 0.25×PBS).
9. EtOH for keeping cell areas, pipettes, etc. sterile.
10. DAPI: 4′-6-diamidino-2-phenylindole (Invitrogen), diluted to 300 nM in PBS.
11. Trypan blue (Sigma) exclusion method, diluted 1:5 in deionized water (see Note 1).
12. Calcein AM 0.5 μM (Invitrogen).

2.3. Single-Cell Seeding

1. Cell scraper (BD Falcon).
2. Flow cytometer with sorting capabilities. All major brand flow cytometers can run the described sorting (they only differ in type of software, protocols, etc.). For more information on cell sorting (into plates), contact your instrument supplier for help. This protocol describes the procedure with a FACS Vantage SE Cell Sorter fitted with a 100 μm nozzle, a plate holder for the motorized *x*/*y* stage, a slide holder to be placed onto the plate holder, and the software CloneCyt Plus (all from BD Biosciences) (see Note 2).
3. FACS clean solution (BD Biosciences).

4. Sterile milli-Q water.
5. 0.25×PBS (see Subheading 2.2, item 8).
6. Humidity chamber for microplate cell culture made of a Petri dish (diam. 100 mm). Any breathable plastic tray will be fine, with the bottom covered with deionized sterile water. To keep the microplate above the water level, two smaller Petri dishes (diam. 20–30 mm, floating on the water), can be used as a plate holder within the larger Petri dish.
7. Absorbing paper tissue that are clean and flat (Precision Wipes, Kimtech Science).

2.4. Detection

1. Light microscope. Here, an Olympus BX51 light microscope outfitted with a manually adjustable *x*/*y*-stage, a 10× objective, and a digital camera (Olympus Camedia C-4000 zoom) was used.
2. Fluorescence microscope. Here, a Zeiss LSM 510 Meta confocal microscope, equipped with a motorized *x*/*y*-stage and a 10× objective, was used.

3. Methods

3.1. Design of the Microwell Plate

1. The microwell plate is constructed as a sandwich structure with three levels: a bottom glass plate (coverslip thickness: 175 μm), an etched silicon microgrid plate and a semipermeable top membrane (Fig. 1a). The resulting microplate is thin (<900 μm including the top membrane) and flat with transparent wells, which makes it useful for imaging and high-resolution microscopy. The microwell plate has the outer format of an array/glass slide (75 × 25 mm) to facilitate implementation in clinical laboratories and standard instruments (Fig. 1b). The microwell plate was developed to fit as many wells as possible while maintaining the possibility to seed individual cells with high precision using flow cytometry.
2. For cell culture in the microwell plate, use the transparent top membrane that seals each well and permits exchange of gases but prevents evaporation of the nanoliter sized wells.

3.2. Manual Cell Seeding (Random Settling of Cells)

The first step of presterilization of the microwell plate is only needed when fabricating your own plates and membranes (see Note 3).

1. A couple of hours (preferably 1 day) before the experiment: Prepare the microwell plates by wrapping it in steam sterilizable plastic bags and seal. Repeat with the cell culture membranes and seal it in an individual bag. One microwell plate can be left unsterilized, to set the sorting positions of the flow cytometer.

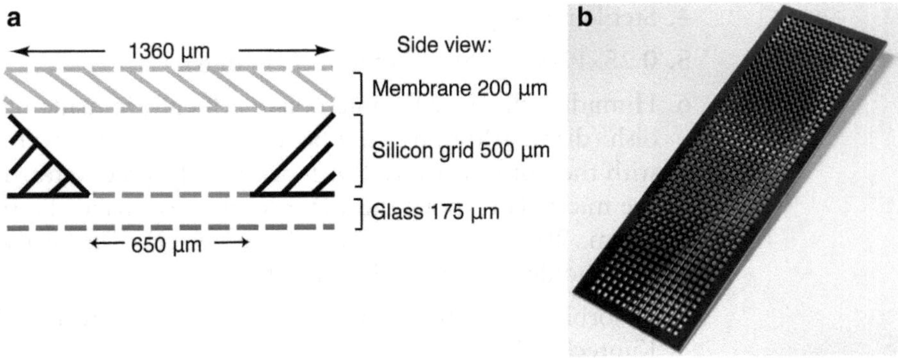

Fig. 1. (**a**) Schematic side view of the microplate. The semipermeable membrane consists of PDMS and is manually added after cell seeding. The KOH etched silicon microgrid creates squared microwells when anodically bonded to the glass bottom. Each microwell has sloped walls with the dimensions 1,360 × 1,360 μm² (*top*) and 650 × 650 μm². The pitch between each well is 140 mm, giving a center–center distance of 1,500 μm between the wells. (**b**) The microwell plate (75 × 25 mm²) holds 672 wells, each with a volume of 0.5 μl.

Autoclave at 121°C for 20 min and let the plates cool to room temperature.

2. It is crucial that all steps below are carried out in a laminar airflow bench to avoid contamination.
3. Prepare the cells in a dilution that is suitable for the experiment. For example, if the goal is to have a single cell in most of the wells, then do the counting of cell concentration as follows:
   (a) One cell per well equals one cell per 0.5 μl (well volume), i.e., 2 cells/μl.
   (b) Dilute your cell sample in culture medium according to (a), meaning 2,000 cells/ml.
   (c) If you on the other hand want ~50 cells per well, you should dilute your cell sample to 100,000 cells/ml.
4. Place the microwell plate on a flat piece of paper tissue, that can adsorb potential excess of liquid (see Note 4). The entire process for manual cell seeding is shown in Fig. 2. For adherent cell growth, pretreat the plate with extracellular matrix (e.g., fibronectin, poly-lysine, etc.) depending of your cells' choice. The protocol for this can be found in Note 14.
5. Add the cell solution over the area of the microwell plate that you wish to study. It is not always necessary to use the full plate: to get more experiments out of one plate, use one area at a time. 200 μL solution will cover approximately 200–300 wells. 800 μL is enough volume to cover the whole plate. Use the pipette tip to scrape the cell solution out to the areas you wish to study/detect.
6. Press the liquid into the wells (and out of the borders of the array, handle it on an adsorbing tissue) with the membrane

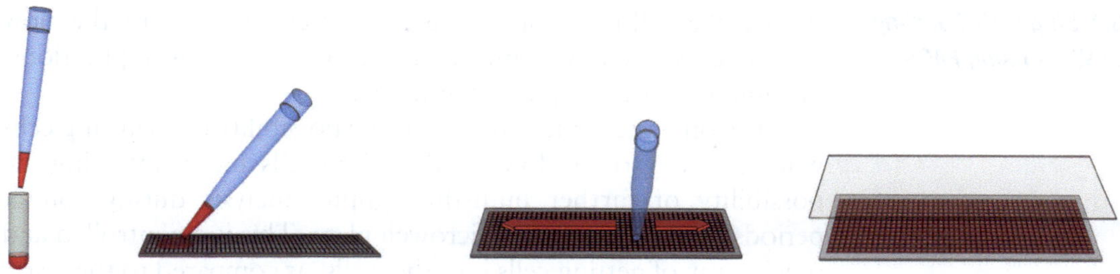

Fig. 2. Schematic description of how cells (or liquid) are spread manually in the microwell plate, using the tip of a pipette. A cell culture membrane is finally placed on top of the plate to seal the wells. Image courtesy by Dr. Fredrik Hellborg.

that you grip with your hands (use sterile gloves!) or using a flat tweezer. If instant imaging is the goal, a coverslip can be placed instead of the membrane. This gives you a thin glass in the bottom as well as on top of the wells.

7. For overnight (or longer) culture, place the plate in the humidity chamber (see Subheading 2.3, item 6) in a standard cell incubator (37°C, 5% CO_2). The individual well-numbering on the plate will help you to keep track of the cells (see Note 5).

3.3. Preparation: Fluorescence Activated Cell Sorting

1. Before a single-cell sorting can be performed, the fluidics of the instruments and the area around it, must be carefully cleaned (see Note 10). FACS Clean solution (or EtOH) followed by sterile water can be flowed through the flow cytometer prior to experiments to obtain sterility (see Note 12).
2. Install the slide holder onto the stage with the plate holder. Make sure that the stage connects and runs with the software.
3. Gating of the cells to be seeded is determined from their individual forward scattering (FSC) and side scattering (SSC) properties (i.e., size and granularity) or by fluorescence. The cells are preferentially sorted at a flow rate of 50–100 cells/s. The flow rate (events/s) can be set as low as possible, while maintaining a stable stream of cells, since the time limiting factor is the movement of the motorized *x*/*y*-stage rather than the flow rate for analysis and/or sorting. In order to study single cells from an entire cell population, all cells within a sample can be selected (gated) for sorting into the microwell plate.
4. Setting the plate-device and start/end-positions: Use the software to program where the instrument will deposit the single cells, by stating how many columns and rows the device have (14 columns and 48 rows for this microwell plate), followed by setting a home position (top left well) and an end position (bottom right well) for coordinates. Use the nonsterilized microwell plate for this type of programming of the FACS instrument (see Note 11).

3.4. Single-Cell Sorting to Wells Using FACS

Before the following steps are performed, make sure the flow cytometer is carefully aligned and adjusted for sorting applications, including correct drop delay settings etc.

A strong asset of the method is the possibility of analyzing cells while they are sorted down in the microwells and maintaining the possibility of further high-throughput analysis during longer periods of time using the microwell plate. This is a controlled and precise way of getting cells into the wells, as compared to the faster but random manual cell seeding protocol in Subheading 3.2.

1. Make sure the settings and gating for your cell sample is good. Sort the cells directly from culture medium or the analysis buffer of your choice. PBS can also be used as a physiological-like buffer. A suitable cell concentration in the test tube is 10^5–10^6 cells/ml and a total volume of 1 ml.
2. During analysis and sorting, sterile 0.25×PBS works fine as sheath buffer.
3. Make sure that the home position of your device matches the position of the microwell plate. Use "test mode" and your test-microplate (nonsterile) for this purpose.
4. Make sure that you have a clean bench (use EtOH) next to the instrument on which you can handle the microwell plate (eventhough it cannot get completely sterile outside the cell hood, it helps a lot to wear gloves and always wipe active surfaces/benches with EtOH).
5. Place the sterile microwell plate on an adsorbing material, e.g., facial paper tissue. For adherent cell growth, pretreat the plate with extracellular matrix (e.g., fibronectin, poly-lysine, etc.) depending of your cells' choice. The protocol for this can be found in Note 14. No pretreatment is required for suspension cells (neither for many adherent cells).
6. Manually dispense culture medium (no cells!) onto the plate (800 μL per whole microwell plate). Distribute the liquid with a sterile cell scraper and remove excess liquid by scraping it over (the edge of) the plate. Each microwell should be compartmentalized from the neighboring well, containing a fixed volume of culture medium.
7. Place the microwell plate in the standard slide holder in a flow cytometer with plate-sorting capabilities.
8. Begin automatic single-cell seeding (time: approximately 1 well/s). The chosen number of cells (e.g., 1 cell) is sorted in each well (Fig. 3, day 1). The wells are already prefilled with growth medium (see step 6 above).
9. When cell seeding is performed, gently pick out the microwell plate and seal with the membrane (see Note 13).

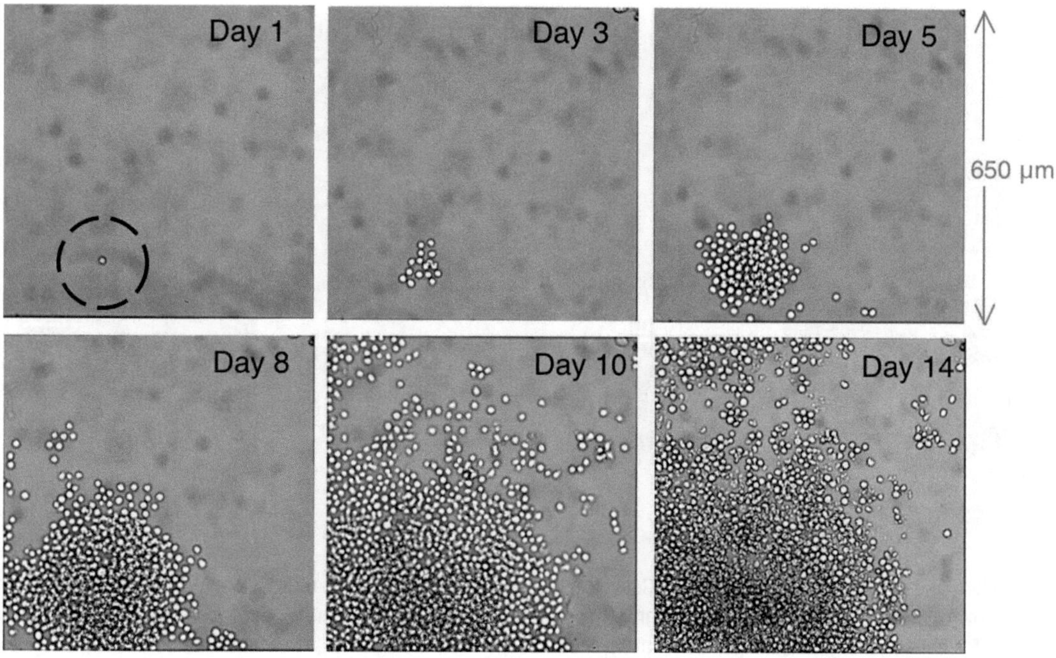

Fig. 3. Proliferation analysis. Clone formation in one well (days 1–14) starting with a single leukemic K-562 cell on day 1.

10. For long-term single cell/clone analysis, place the plate in a humidified atmosphere (for example a humidity chamber as described above) in a standard cell incubator, in 37°C, 5% CO_2.

3.5. Long-Term Cell Culture and Post-analysis by Fluorescence Staining

When the cells have been placed in the wells, either by manual seeding or by controlled automatic cell sorting, they can be incubated and grown for several weeks (this is of course dependent on the cell type). With the use of a microscope, one can study how the cells expand over time, forming clones (Fig. 3, day 3–14) as they grow. The entire method is outlined in Fig. 4.

As a final step, the cells can be post-analyzed (Fig. 3d) by fluorescence markers. Below are three ways of staining described, but any type of reagents can of course be added in a similar manner (e.g., to study different drug concentrations on different cells, investigating peptides, etc).

1. This protocol can be applied to cells that grow adherent onto the glass bottom of the wells.
2. Remove the growth medium/analysis buffer from the wells by rapidly turning the plate upside down on a clean tissue, followed by placing it upright again (see Note 8).

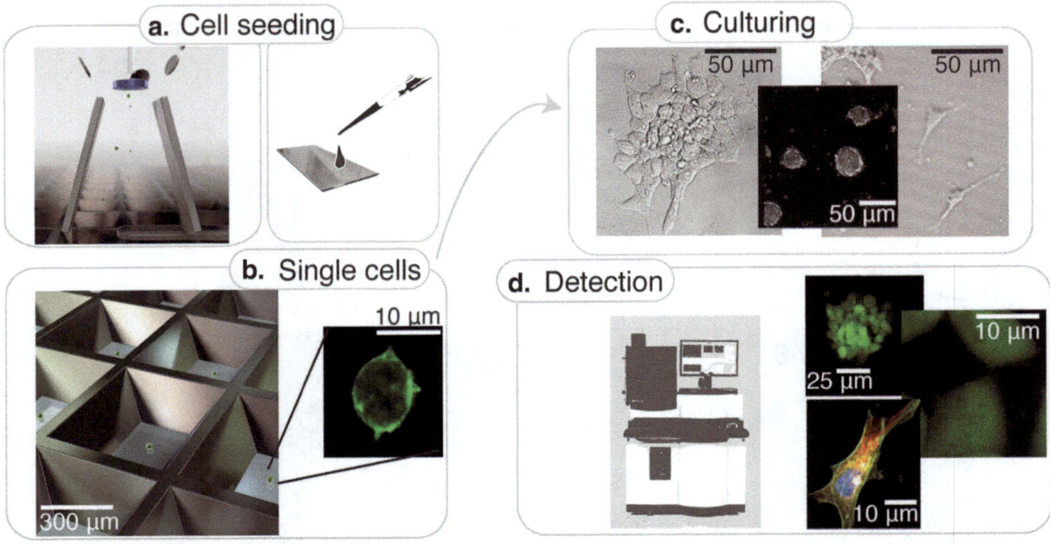

Fig. 4. Schematic overview of the method. (**a**) Cells are seeded into the microwells, either by automatic instrumentation such as flow cytometric cell-sorting (*left*), or manually by limited dilution in a random manner (*right*). (**b**) One cell per well opens up for heterogeneity screenings, clonal assays, etc. Zoom in on a single cell, fixed and labeled directly after cell seeding. (**c**) Cell analysis, weeklong culturing and differentiation studies can be performed. Culture medium change can be performed, by rinsing the chip with fresh medium. (**d**) The entire microwell plate can be screened using conventional automated imaging systems, detecting cells and clones in the 672 individual wells in a rapid manner.

3. Add your fluorescence marker or staining reagent of choice (according to the manufacturer's instructions). Three examples are described below:
   (a) The nuclear probe DAPI is used for staining of the nuclei. Incubate 5 min.
   (b) The vital stain Trypan blue is used to selectively color dead tissues or cells blue. Incubate 5 min.
   (c) Calcein AM is transported through the cellular membrane into live cells, which makes it useful for testing of cell viability and for short-term labeling of cells. Incubate 5 min.
4. For primary and secondary antibody staining, follow the same principle as above, but with other incubation times etc. This procedure is described in more detail elsewhere (10).

Other detection methods (array scanners, plate readers) can also be used instead of, or complimentary to, microscopic detection.

4. Notes

1. Due to the small volumes of each microwell (0.5 µl) it will be difficult to visualize the unlabelled living cells with the darkblue Trypan dye. (Bright-field microscopy can be used for detection).

Therefore, the Trypan blue dye was prediluted 1:5 in deionized water.

2. For cell sorting, these are the standard tools that usually comes with the instrument. No additional accessories should be needed, meaning that if you can do cell sorting into 96-well plates, you can also do the 672-well sorting as described here.
3. If you buy the microwell plate and the culture membrane, they are already sterile and ready-to-use.
4. Make sure that you use clean and sterile hands, tweezers, and paper tissues, to prevent contamination when loading the plate with cells and/or placing the culture membrane on top of the plate.
5. Each well is numbered (column:row) to help keep track of the cells. The numbers are placed on top of the silicon walls and are visualized using a microscope.
6. For optimal transparency and high-magnification imaging, place a coverslip on top of the plate (instead of the culture membrane). By doing so, you have glasses of coverslip thickness both in the bottom and top of each well.
7. The volume of the wells (0.5 μL per well) is big enough to handle with a standard pipette (0.1–2.5 μL or 0.5–10 μL). There are also low-volume pipettes available with high precision if wanted.
8. Removal of liquid from the plate can be solved by for example; (1) turning the plate upside down on a clean tissue, (2) evaporation of liquid, or (3) aspiration of liquid. The choice depends on the assay being run.
9. Rinsing and/or dipping the plate in a beaker can be suitable for washing. The sloped walls and shallow wells will facilitate liquid exchange, i.e., the plate can be treated as a standard glass slide.
10. Be sure to maintain proper cell handling techniques when working with cells, to avoid contaminations. A single bacterium can destroy an entire experiment, since the well-volume is only 0.5 μl, and upon contamination a well will soon be full of rapidly dividing bacteria (instead of the cells of your interest).
11. When adjusting the first (home) and last (end) dispensing positions in the FACS instrument, place a plastic tape on top of the plate and you will easily see the exact position of the dispensed droplets. (By doing so, you prevent the drops to enter the wells and thus become invisible for your eyes).
12. Depending on the cells that you wish to sort, you can choose the nozzle size of your flow cytometer. Make sure to install this before starting the cleaning procedure.
13. Placing the membrane on top of the plate: To avoid air bubble formation, one to two drops of culture medium can be added on top of the plate to facilitate plate sealing with the culture

membrane. The excess of fluid will evenly spread under the membrane.

14. For coating of the wells using chosen ECM reagent, follow this protocol:
    (a) Work in a laminar airflow bench.
    (b) Place the microwell plate onto an adsorbing flat facial tissue.
    (c) Pipette 800 μl of your coating material of choice (concentrations, temperature, timing, etc. according to the manufacturer's recommendations) onto the microwell plate. Spread the liquid into all wells using the pipette tip (same procedure as in Fig. 2) and incubate.
    (d) If needed (depends on the coating reagent), rinse the plate with washing buffer (preferentially the growth medium/ analysis buffer that you will use later on when placing the cells in the wells).
    (e) If needed (depends on which assay that will be run), empty the wells according to Note 8.

References

1. Spudich JL & Koshland DE, Jr. (1976) Non-genetic individuality: chance in the single cell. *Nature* 262(5568):467–471.
2. Templer RH & Ces O (2008) New frontiers in single-cell analysis. *Journal of the Royal Society, Interface/the Royal Society* 5 Suppl 2:S111-112.
3. Lindstrom S & Andersson-Svahn H (2011) Miniaturization of biological assays – overview on microwell devices for single-cell analyses. *Biochim Biophys Acta* 1810(3):308–316.
4. Di Carlo D & Lee LP (2006) Dynamic Single Cell Analysis for Quantitative Biology. *Analytical chemistry* 78(23):7918–7925.
5. Longo D & Hasty J (2006) Dynamics of single-cell gene expression. *Molecular systems biology* 2:64.
6. Sims CE & Allbritton NL (2007) Analysis of single mammalian cells on-chip. *Lab on a chip* 7(4):423–440.
7. Voldman J (2006) Engineered systems for the physical manipulation of single cells. *Current opinion in biotechnology* 17(5):532–537.
8. Lindstrom S & Andersson-Svahn H (2010) Overview of single-cell analyses: microdevices and applications. *Lab on a chip* 10(24): 3363–3372.
9. Lindstrom S, Larsson R, & Svahn HA (2008) Towards high-throughput single cell/clone cultivation and analysis. *Electrophoresis* 29(6):1219–1227.
10. Lindstrom S, *et al.* (2009) High-Density Microwell Chip for Culture and Analysis of Stem Cells. *PloS one* 4(9):e6997.
11. Taylor LC & Walt DR (2000) Application of high-density optical microwell arrays in a live-cell biosensing system. *Analytical biochemistry* 278(2):132–142.
12. Chin VI, *et al.* (2004) Microfabricated platform for studying stem cell fates. *Biotechnology and bioengineering* 88(3):399–415.
13. Revzin A, Sekine K, Sin A, Tompkins RG, & Toner M (2005) Development of a microfabricated cytometry platform for characterization and sorting of individual leukocytes. *Lab on a chip* 5(1):30–37.
14. Rettig JR & Folch A (2005) Large-scale single-cell trapping and imaging using microwell arrays. *Analytical chemistry* 77(17): 5628–5634.
15. Deutsch M, *et al.* (2006) A novel miniature cell retainer for correlative high-content analysis of individual untethered non-adherent cells. *Lab on a chip* 6(8):995–1000.
16. Tokimitsu Y, *et al.* (2007) Single lymphocyte analysis with a microwell array chip. *Cytometry A* 71(12):1003–1010.

Chapter 6

Isodielectric Separation and Analysis of Cells

Michael D. Vahey and Joel Voldman

Abstract

Measuring the electrical properties of a cell provides a fast and accessible means of identifying or characterizing cells whose biological state differs from the population as a whole. This chapter describes a microfluidic method for characterizing the electrical properties of cells based upon their convergence to equilibrium in an electrical conductivity gradient. The method, called isodielectric separation, uses the dielectrophoretic force induced on polarizable objects in spatially nonuniform electric fields to deflect cells to the point in the conductivity gradient where their polarization charge vanishes. This equilibrium position encodes the cell's electrical properties and can be used to identify cells that are electrically distinct from a background population, to determine the extent of this difference, and to physically isolate them for further study.

Key words: Electrical separation, Dielectrophoresis, Cell separation, Electrical analysis, Microfluidics

1. Introduction

The intrinsic physical properties of cells have the potential to encode valuable biological information. A wide variety of techniques have been developed to gain access to this information, characterizing and sorting cells according to differences in their size, density, rigidity, impedance, or other physical properties (1–9). The efficacy of these techniques depends on three characteristics of the intrinsic property on which they are based: (1) the property should be easily measurable at a single-cell level; (2) variations in the property should correlate with biologically relevant variations; and (3) it should be possible to identify and isolate cells possessing these variations by some means. One area which has proven to be particularly well suited to satisfying these requirements is electrical methods: techniques designed to characterize and separate cells based upon their intrinsic electrical properties (i.e., their conductivity and permittivity). Developments in microfabrication over the past

Sara Lindström and Helene Andersson-Svahn (eds.), *Single-Cell Analysis: Methods and Protocols*,
Methods in Molecular Biology, vol. 853, DOI 10.1007/978-1-61779-567-1_6, © Springer Science+Business Media, LLC 2012

few decades have made the integration of cell-sized electrodes into multifunctional devices routine, and because the electrical properties of a cell depend intimately on the cell's structure and composition, differences in these properties are often biologically relevant. These features have established electric fields as an important means of characterizing, sorting, and manipulating populations of cells down to the single-cell level.

Although electrical methods are tremendously diverse, two widely used categories are those based on measuring the impedance of cells, and those based upon a cell's dielectrophoretic response. While impedance-based methods measure the change in resistance and capacitance between electrodes induced by the presence of one or more cells (10–12), dielectrophoretic methods leverage the force exerted on a polarizable object in the presence of spatially nonuniform electric fields (13). The direction of the dielectrophoretic force depends on the electrical properties of the cell relative to the medium in which it is suspended; a cell with positive net polarizability (e.g., one with higher effective conductivity than its surroundings) will move in the direction of increasing electric field intensity, while a cell with negative net polarizability will move in the opposite direction.

In the characterization and separation of cells, dielectrophoretic methods present an advantage in their ability to use the electrical polarization of cells to both interrogate their properties (specifically, the magnitude and sign of their polarization), as well as to physically move them. This ability to simultaneously interrogate and position cells is one of the more attractive features of dielectrophoretic methods. Unlike methods in which interrogation must be coupled to downstream separation (e.g., impedance cytometry or fluorescence activated cell sorting), dielectrophoresis essentially combines the two steps, leading to potentially faster screens that are further simplified by not requiring cell labeling. We have leveraged these advantages of electrical methods and the ability to position cells in different locations according to differences in their electrical properties by developing a new separation and characterization method called isodielectric separation (IDS) (14). In IDS, cells and particles are dielectrophoretically concentrated to the regions in an electrical conductivity gradient where their polarization charge and the resulting DEP force vanish. Using IDS, we have been able to sort and characterize cells and particles spanning three orders of magnitude in volume and electrical conductivity (15).

Figure 1 illustrates the concept of this method. We create a monotonic gradient in electrical conductivity across the width of a microfluidic channel by injecting one solution of relatively high conductivity containing the cell mixture and a second solution of relatively low conductivity into a device with a diffusive mixer (Fig. 1, left). This mixer generates a smooth monotonic conductivity profile that flows directly into a channel containing electrodes arranged across the diagonal (Fig. 1, right). These electrodes guide the cells

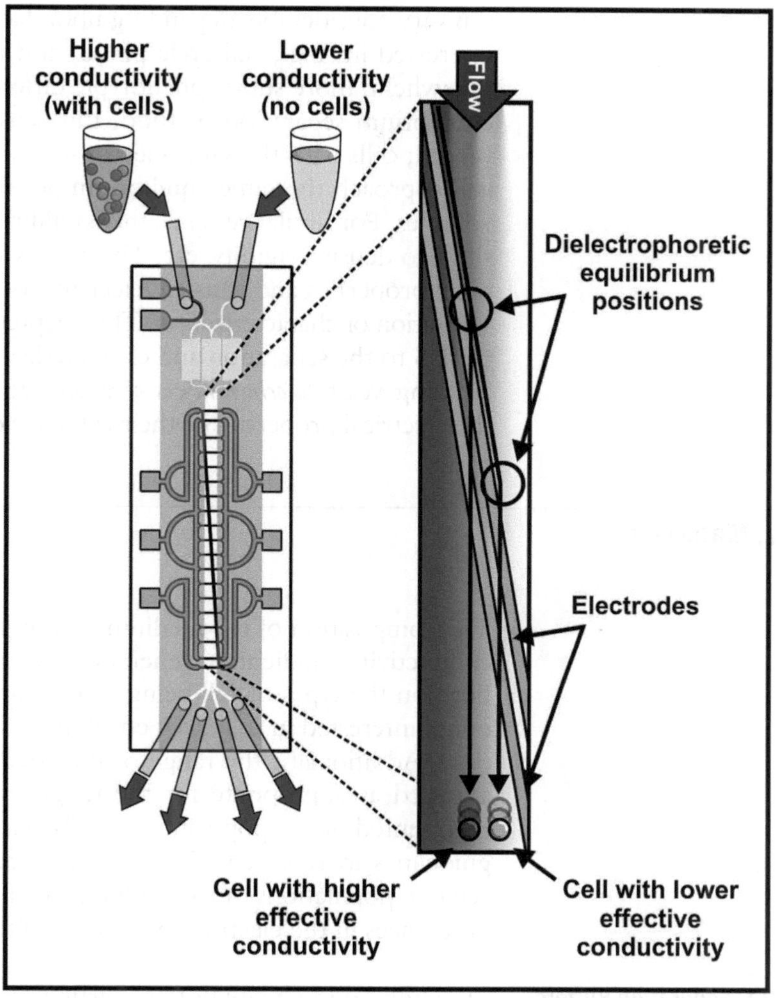

Fig. 1. Analytical separations using IDS. The device is loaded with liquid of higher and lower electrical conductivities, with cells suspended in the higher conductivity. These liquids establish a conductivity gradient across the width of a separation channel, containing electrodes across its diagonal. The cells flow through this channel and are dielectrophoretically deflected by the electrodes across the conductivity gradient until they reach their dielectrophoretic equilibrium positions, where they pass over the electrode barrier. Finally, the cells or particles flow to an observation region, where the cells and particles are imaged to determine their spatial distributions, and to outlets, where fractionated samples may be collected.

in the direction of decreasing medium conductivity—a one-sided approach to equilibrium—until the DEP force becomes sufficiently small that it is overwhelmed by hydrodynamic drag and the barrier is breached. The cells then continue downstream unobstructed for collection. Sampling cells from different portions along the channel width thus segregates cells according to their electrical properties.

Although a wide variety of dielectrophoretic techniques for characterizing and sorting cells have been developed, a common limitation of these approaches is their high sensitivity to the size of a cell. While for some applications this may be desirable, cell size can vary considerably depending upon factors that one may not be interested in (e.g., cell cycle phase) and thus has the potential to overwhelm more subtle phenotypic differences. Because IDS is an equilibrium separation method, this sensitivity to size is circumvented; cells with the same electrical conductivity and permittivity will approach the same equilibrium position independent of their volumes. For similar reasons, the equilibrium position is also insensitive to density, rigidity, etc. This makes IDS specific only to electrical properties and thus an excellent tool for either electrical cell separation or characterization. This chapter describes the application of IDS to the separation and characterization of populations of the budding yeast *Saccharomyces cerevisiae* according to differences in the electrical properties of their cell envelope.

2. Materials

The composition of the medium (i.e., liquid) used to establish the conductivity gradient is generally application specific, depending both on the types of cell being studied and the parts of these cells one is interested in (e.g., cell envelope, cell membrane, or cell interior). Additionally, the range of the conductivity gradient may be adjusted as appropriate for the range of electrical conductivities represented by the population of cells being studied. The following guidelines are representative of those we have used to sort and characterize populations of the budding yeast *S. cerevisiae* according to differences in the electrical properties of their cell envelope.

2.1. Separation Buffers

1. High-conductivity buffer: starting with 40 ml deionized water (conductivity of $\sim 5.6 \times 10^{-8}$ S/m at room temperature), add 1.5 ml phosphate-buffered saline (conductivity of ~1.5 S/m at room temperature; Gibco, Carlsbad, CA) and 0.2 ml bovine serum albumin for the higher conductivity separation buffer (conductivity of ~0.065 S/m).
2. Low-conductivity buffer: for the lower conductivity buffer, add 0.2 ml bovine serum albumin to 40 ml deionized water (conductivity of ~0.01 S/m). The intermediate conductivity can be achieved by mixing these solutions 1:1. For mammalian cells, osmotically balanced solutions can be prepared by replacing deionized water with a solution of deionized water containing glucose at a concentration of 300 mM.

2.2. Device Fabrication

1. Clean-room facilities capable of photolithography and metal deposition.
2. Six inch Pyrex wafers and silicon wafers (Bullen, Eaton, OH).

3. SU8 2015 (MicroChem, Newton, MA) and NR7-3000P (Futurrex, Franklin, NJ) photoresists and developers.
4. Poly(dimethylsiloxane) (Dow Corning, Midland, MI).
5. Customized printed circuit board (ExpressPCB).

2.3. Additional Equipment and Instrumentation

1. Three glass luer-lock syringes (1000 series, Hamilton Company, Reno, NV).
2. PEEK tubing (1561), luer adapters (P-618, P-135), ferrules (P-200), and fittings (P-235); (all available from IDEX Health and Sciences, Oak Harbor, WA).
3. Syringe pump (KD Scientific 200, Holliston, MA).
4. Function generator (33220A, Agilent, Palo Alto, CA).
5. Fluorescence microscope with 5× objective and camera.

3. Methods

3.1. Device Fabrication and Assembly

The device consists of two parts: a microfluidic channel that encloses the liquid and the cells, and which is made of PDMS cast onto an SU-8 mold, and a glass substrate containing patterned electrodes. Although processing parameters may vary for different facilities, those listed here provide general guidelines. Consult the MSDS and follow appropriate protocols for the use and disposal for of all chemicals listed here.

Molds for microfluidic channels

1. Dehydration bake the clean silicon wafer(s) on a hot plate at ~200°C for ~30 min. Dispense ~6 ml of SU8-2015 photoresist on the center of the wafer and ramp to 500 rpm at 100 rpm/s, hold for 5–10 s, then ramp to 2,250 rpm at 300 rpm/s and hold for 30 s. This achieves a film thickness of ~20 μm, defining the depth of the microfluidic channel (see Note 1).
2. Prebake the wafer(s) using a slow ramp on a hot plate from 60 to 95°C, hold at 95°C for 2 min; then allow the wafer to cool to room temperature. UV expose the wafer through the flow chamber mask, using a total dosage of ~120 mJ/cm^2 at 365 nm. For the postbake, repeat the parameters of the prebake (see Note 2).
3. Develop the wafers for 3–5 min using PM acetate. To remove any residual developer and to dry the wafer afterwards, we perform a 30 s spin while spraying with PM acetate, a 30 s spin while spraying with isopropanol, and a 30-s spin dry (see Note 3).
4. Before applying PDMS to the wafer to mold devices, the wafer should be silanized; place it in a vacuum chamber along with

three to four drops of HMDS and allow it to set (with vacuum on) for approximately 30 min (see Note 4).

Electrodes

1. Perform a dehydration bake on the clean Pyrex wafer(s) at ~120°C for ~30 min and apply photoresist adhesion promoter (HMDS).
2. Coat the wafer(s) with a 1–2 μm layer of NR7-3000P negative photoresist (Futurrex). For a 6 in. wafer, we allow the resist to spread for 6 s at 750 rpm, followed by a 30-s spin at 2,500 rpm.
3. Prebake wafer(s) on a hot plate at 155°C for 90 s. UV expose the wafer through the electrode mask. The UV dosage should be optimized; we use an exposure energy of ~100 mJ/cm^2 at a wavelength of 365 nm. Bake (on a hotplate) at 120°C for 2 min.
4. Gently agitate the wafer(s) in Resist Developer RD6 (Futurrex) for 25–30 s to develop, then rinse with deionized water and spin dry.
5. Deposit metal layers (100 Å Ti, 2,000 Å Au) on the resist-patterned wafers and then immerse them in acetone to lift off the metal film overlying the photoresist (see Note 5).
6. Dice the wafers to separate the individual dies.

Device assembly

1. Create the flow chamber by casting PDMS (10:1 base to curing agent ratio) on the patterned silicon wafer to a thickness of ~4 mm and allowing it to cure (see Note 6).
2. Remove the cured PDMS and carefully cut out a chamber. Punch holes for each of the inlets and outlets (0.06 in. diameter) so that tubing can be press-fit into the device to load and collect samples (see Note 7).
3. Align the chamber to an electrode chip and bond by exposing the contacting surface to oxygen plasma for ~1 min (see Note 8). Form electrical contacts to the chip, and seal around the bonded PDMS with an insulating epoxy to prevent leaking (see Note 9).

3.2. Sample Preparation and Loading

1. Before preparing or loading the cells, prime the device using deionized water. To do this, connect tubing to one of the inlets while leaving the other inlets and outlets open. Manually driving liquid through the device will cause it to fill and leave droplets at the inlets and outlets. Continue to perfuse the device with water until any internal bubbles have been removed (see Note 10).
2. Wash the sample in the highest conductivity separation buffer, and adjust the cells to the desired concentration (see Note 11). If visual characterization of the cell's equilibrium positions is

desired, they can be fluorescently labeled; this should be done prior to washing and resuspending cells in the high conductivity medium.

3. Load the cell suspension and the lower conductivity separation buffers into separate glass syringes, removing any bubbles from the syringes. Using a luer-lock adapter, attach tubing to the syringes; place the syringes into the pump and activate the pump so that liquid flows through each syringe at a rate of ~5 μl/min. Once the tubing is filled with liquid (indicated by small droplets forming at the ends of the tubing), insert the tubing into the device (see Note 12).
4. Verify (using a microscope) that the sample is entering the device with the high conductivity buffer, and that the fraction of the channel containing the lower conductivities does not contain any cells. If bubbles are present, allow the device to run until they have been removed (see Note 13). Once the sample loading appears to be steady, you are ready to reduce the flow rate and begin the experiment. A total flow rate (i.e., combined flow through all syringes) of 4.5 μl/min (±50%) is appropriate for most applications (see Note 14).
5. Connect the function generator to the leads of the PCB and adjust the settings to appropriate values for your sample. For *S. cerevisiae* in a conductivity gradient spanning 0.01–0.065 S/m, the decade of frequencies from 0.1 to 1 MHz should generally be appropriate (see Note 15).

3.3. Sample Characterization and Collection

1. Visual characterization of a sample consists of recording the position along the channel width where cells localize. For example, if the sample contains two differently labeled populations, the positions at which these samples pass through the electrode barrier gives their effective conductivities relative to each other at the particular electric field frequency being used. Figure 2 illustrates a typical experimental setup for this type of experiment.
2. Spectral characterization is performed by recording shifts in the equilibrium position of cells as the frequency is varied (see Note 16). The general concept behind characterizing labeled populations by varying the operating conditions is illustrated in Fig. 3.
3. In many applications, it is not practical (or possible) to label different populations for imaging, and it is necessary to collect the sample for follow-up studies. The sorted fractions of cells may be collected directly from the device's outlets in a variety of ways, depending upon the volume of the sample and the number of cells that are needed (see Note 17).

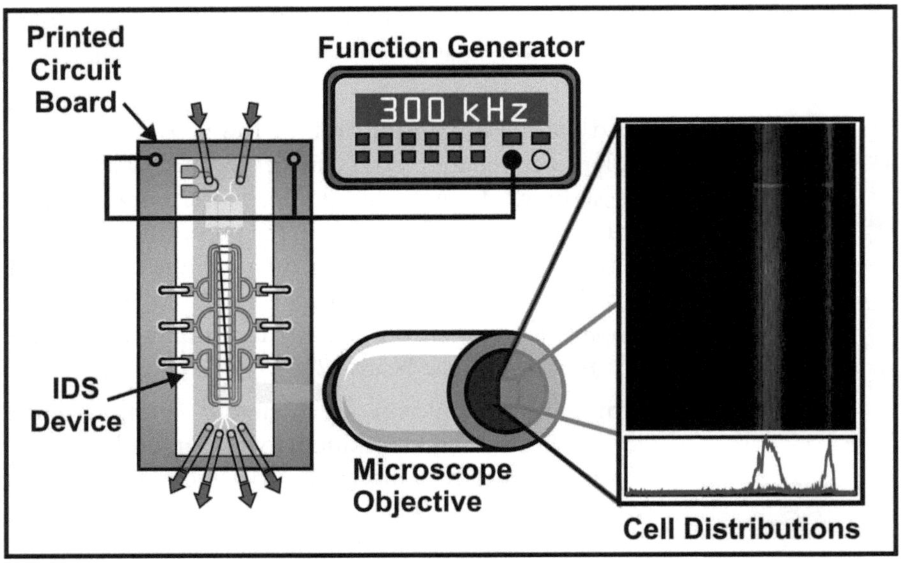

Fig. 2. A schematic layout of the device and accompanying instrumentation.

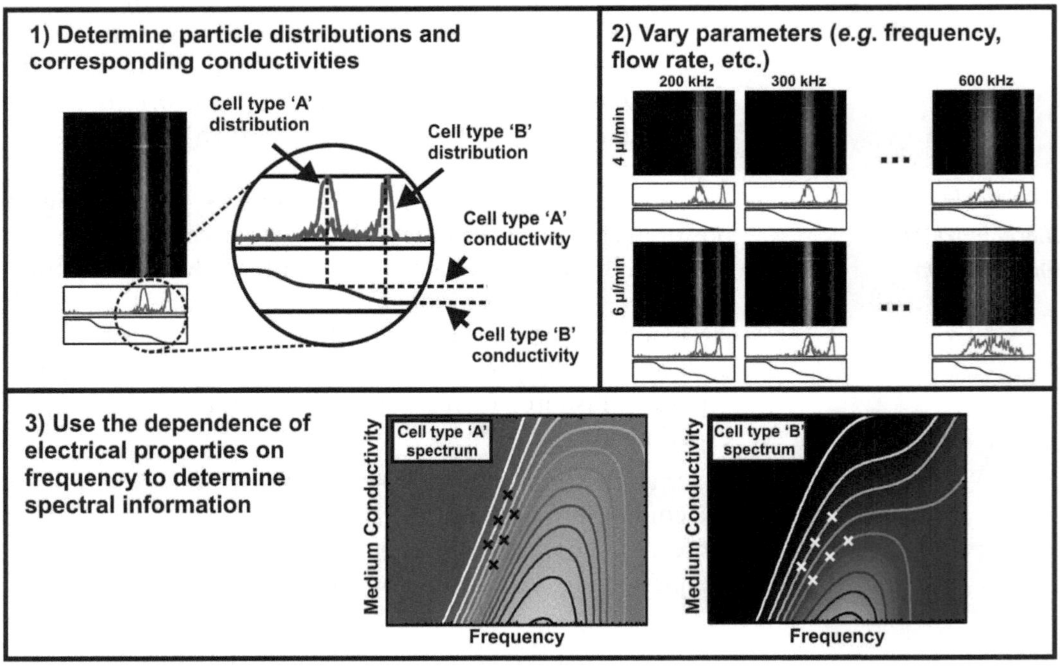

Fig. 3. Typical process for characterizing two populations of cells. Two populations of cells ("A" and "B") are fluorescently labeled and imaged as they pass through the IDS device. The spatial distributions of these cells give the conductivity at which they pass through the dielectrophoretic barrier presented by the electrodes (step 1 above). Varying operating parameters such as frequency, voltage, and flow rate while observing changes in the equilibrium conductivity of the two cell types determine the polarization spectrum of the different cell types (steps 2 and 3 above).

4. Once the samples are collected, they are amenable to characterization by any standard molecular or cell biology technique. For example, collecting and counting cells from the outlets provides an alternate means of determining their distribution across conductivity that does not require microscopy. Alternatively, if the population contains genetic variants (or is clonal but exhibits heterogeneous gene expression), the relative abundance of different strains in each sorted fraction can be determined through quantitative PCR, microarray analysis, or sequencing.

3.4. Cleaning the Device

1. After the experiment, the device can generally be cleaned and reused. Rinse the device thoroughly with deionized water; this can follow a rinse with detergent in extreme cases, or can be accompanied by sonication.
2. After rinsing with water, dry the device thoroughly by connecting it to a low pressure supply of nitrogen gas.
3. If the device has become clogged or it is not possible to clean it by rinsing or sonication, the device can be taken apart, the PDMS channel removed, and the electrode chip reused. Solvents such as acetone can generally be used to soften the epoxies used during device assembly, and cured PDMS can be dissolved from the electrode chip (Dynasolve 220; Dynaloy, Indianapolis, IN).

4. Notes

1. Dispense the resist slowly, being careful to prevent any bubbles from forming on the wafer.
2. Do not allow the wafer to cool too rapidly. Also, if the prebake is not sufficiently low, the wafer may adhere to the mask during exposure. If this occurs, extend the prebake.
3. We find that using fresh photoresist (i.e., not approaching the expiration date) it critical in obtaining smooth features that do not easily delaminate from the wafer.
4. Once an SU8 mold has been created, it is possible to create many replicas by recasting the mold in durable plastic (16).
5. In some cases where the features are particularly small or dense, the lift-off process can be accelerated by gently swabbing the wafers with an applicator soaked in acetone.
6. It is important that the PDMS layer be neither too thick nor too thin; if too thick, it may be difficult to image the sample using any microscope objective without a long working distance.

If too thin, it may be difficult to press-fit inlet tubing into the device without the sample or separation buffers leaking.

7. Tubing with smaller inner diameters minimizes the residence time of cells during loading, and will improve the uniformity of the loading concentration.

8. Manual alignment of the chip with accuracy better than ~0.5 mm may be difficult. If this is the case, an automated stage can be used for more precise alignment.

9. We find that custom-designed printed circuit boards and conductive epoxy (CW2400, Chemtronics, Kennesaw, GA) are an effective way to establish electrical connections to the chip.

10. Leaving liquid droplets at the inlets of the device after it has been primed provides an interface for inserting the tubing later without introducing any bubbles.

11. Because of electrostatic and hydrodynamic interactions between cells as they pass through the device, the performance will depend on the concentration at which cells are loaded. For the purpose of characterizing cells at a single-cell level, low concentrations will work best. However, for preparative separations (i.e., those in which appreciable numbers of cells are to be collected) higher concentrations can not only improve the throughput, but the sensitivity of the separation as well. In working with *S. cerevisiae*, we find that concentrations between 1 and 5×10^7 cells/ml work well.

12. It is easiest to insert the tubing without introducing bubbles if there are droplets remaining over the inlets from the priming step (see Subheading 3.2).

13. Because PDMS is permeable to gas, increasing the pressure within the channel can force the removal of bubbles through the channel ceiling and walls. This can be achieved by increasing the flow rate, or by plugging the outlet channel.

14. During long-term operation with cells that sediment rapidly, we recommend mounting the syringe pump vertically and making the tubing connecting the syringes to the device as short as possible.

15. Consideration must be given to the operating conditions used in any experiment to prevent fouling of the device. High current densities, even for a very brief time, may irreparably damage the electrodes; as a general guideline, we find that current densities lower than 10^5 A/m^2 do not damage the electrodes.

16. Device fouling can also result if the frequency is set to a value where the cells have strong positive polarizability. When this is the case, they will be attracted to the surface of the electrodes, where they may become stuck and produce clogging. Mechanical

agitation of the device can be effective in freeing cells that have become stuck in this way.

17. We like to use pipette tips, press-fit directly into the device's outlets. These act as reservoirs, filling with sample that can easily be transferred to other containers with a manual pipettor.

References

1. Rosenbluth MJ, Lam WA, Fletcher DA (2008) Analyzing cell mechanics in hematologic diseases with microfluidic biophysical flow cytometry. Lab Chip **8**(7): 1062–70
2. Thevoz P et al (2010) Acoustophoretic synchronization of mammalian cells in microchannels. Anal Chem **82**(7): 3094–8
3. Becker FF et al (1995) Separation of human breast cancer cells from blood by differential dielectric affinity. Proc Natl Acad Sci USA **92**(3): 860–4
4. Chowdhury F et al (2009) Material properties of the cell dictate stress-induced spreading and differentiation in embryonic stem cells. Nat Mater **9**(1): 82–8
5. Cross SE et al (2007) Nanomechanical analysis of cells from cancer patients. Nat Nanotechnol **2**(12): 780–3
6. Dustin ML, Cooper JA (2000) The immunological synapse and the actin cytoskeleton: molecular hardware for T cell signaling. Nat Immunol **1**(1): 23–9
7. Flanagan LA et al (2008) Unique dielectric properties distinguish stem cells and their differentiated progeny. Stem Cells **26**(3): 656–65
8. Pethig R et al (2002) Dielectrophoretic studies of the activation of human T lymphocytes using a newly developed cell profiling system. Electrophoresis **23**(13): 2057–63
9. Suresh S (2007) Biomechanics and biophysics of cancer cells. Acta Biomater, **3**(4): 413–38
10. Gawad S et al (2007) Impedance spectroscopy using maximum length sequences: application to single cell analysis. Rev Sci Instrum, **78**(5): 054301
11. Gawad S, Schild L, Renaud PH (2001) Micromachined impedance spectroscopy flow cytometer for cell analysis and particle sizing. Lab Chip **1**(1): 76–82
12. Holmes D et al (2009) Leukocyte analysis and differentiation using high speed microfluidic single cell impedance cytometry. Lab Chip, **9**(20): 2881–9
13. Pohl HA, Crane JS (1971) Dielectrophoresis of cells. Biophys J, **11**(9): 11–27
14. Vahey MD, Voldman J (2008) An equilibrium method for continuous-flow cell sorting using dielectrophoresis. Anal Chem **80**(9): 3135–43
15. Vahey MD, Voldman J (2009) High-throughput cell and particle characterization using isodielectric separation. Anal Chem **81**(7): p. 2446–55
16. Desai SP, Freeman DM, Voldman J (2009) Plastic masters-rigid templates for soft lithography. Lab Chip **9**(11): 1631–7

Chapter 7

Single Cell Electroporation Using Microfluidic Devices

Séverine Le Gac and Albert van den Berg

Abstract

Electroporation is a powerful technique to increase the permeability of cell membranes and subsequently introduce foreign materials into cells. Pores are created in the cell membrane upon application of an electric field (kV/cm). Most applications employ bulk electroporation, at the scale of 1 mL of cells (ca. one million cells). However, recent progresses have shown the interest to miniaturize the technique to a single cell. Single cell electroporation is achieved either using microelectrodes which are placed in close vicinity to one cell, or in a microfluidic format. We focus here on this second approach, where individual cells are trapped in micrometer-size structures within a microchip, exposed in situ to a high electric field and loaded with either a dye (proof-of-principle experiments) or a plasmid. Specifically, we present one device that includes an array of independent electroporation sites for customized and successive poration of nine cells. The different steps of the single cell electroporation protocol are detailed including cell sample preparation, cell trapping, actual cell poration and on-chip detection of pore formation. Electroporation is illustrated here with the transport of dyes through the plasma membrane, the transfection of cells with GFP-encoding plasmids, and the study of the ERK1 signaling pathway using a GFP–ERK1 protein construct expressed by the cells after their transfection with the corresponding plasmid. This last example highlights the power of microfluidics with the implementation of various steps of a process (cell poration, culture, imaging) performed at the single cell level, on a single device.

Key words: Gene transfection, Gene therapy, Drug delivery, Lab on a chip, Electropermeabilization, Single cell analysis

1. Introduction

The introduction of foreign materials in cells is essential for a great variety of applications in medicine and biotechnology such as gene therapy, drug delivery, cell engineering, or microorganism inactivation. Different techniques that rely on various chemical or physical principles exist to cross the impermeable barrier formed by biological membranes and to transiently permeabilize them. One of the most popular techniques is electroporation (1), which is based on the use of an electric field. Upon application of short pulses (exponentially

Sara Lindström and Helene Andersson-Svahn (eds.), *Single-Cell Analysis: Methods and Protocols*, Methods in Molecular Biology, vol. 853, DOI 10.1007/978-1-61779-567-1_7, © Springer Science+Business Media, LLC 2012

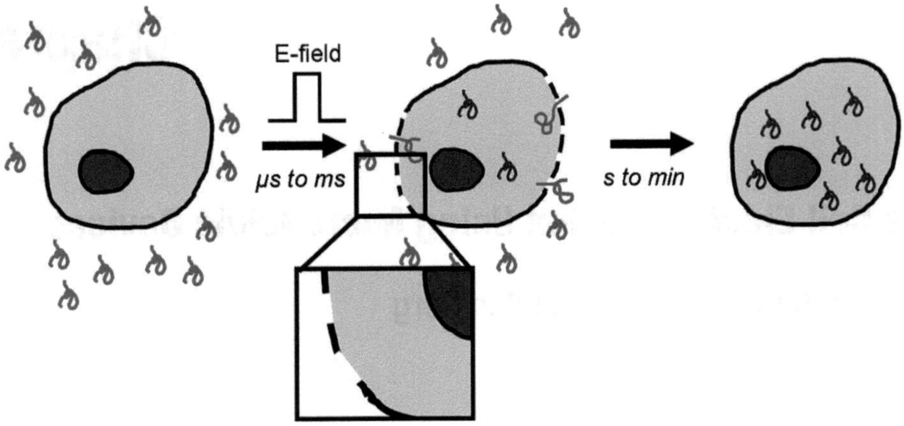

Fig. 1. Principle of cell electroporation. *From left to right*: A cell is incubated in a medium supplemented with foreign substances to be introduced in the cell. The cell is exposed to an electrical treatment consisting of a short pulse (μm to ms) of a high electric field (kV/cm); this leads to the formation of pores in the cell plasma membrane, enabling the entry of foreign substances inside the cell. After a few minutes, the pores seal again, leaving cells loaded with foreign materials (Courtesy of Dr. I. van Uitert).

decaying pulses or square pulses) of a high electric field (kV/cm range) the cell transmembrane potential increases. If the latter reaches the critical threshold of 0.2–1 V (2), a transient molecular rearrangement of the phospholipids forming the membrane occurs, which leads to the creation of aqueous pores in the membrane in a few microseconds, as illustrated in Fig. 1. These pores consist of pathways for molecular exchanges between the intra- and the extra-cellular media. If the electrical protocol remains mild, pores reseal within seconds to minutes (3), and the cells survive the treatment after having taken up foreign materials added in the solution. However, this viable cell poration is just one of the three scenarios observed when a cell is exposed to an electric field. Alternatively, when the electric field is too weak, the cells do not appear to be affected by the treatment: pores are too small or nonexistent. Conversely, if the electrical protocol is too strong, cells undergo lysis and die. Unfortunately, little is known about the process of pore formation in the cell membrane, and these three possible scenarios are difficult to predict to yield a safe and efficient methodology. What is now acknowledged is that pore formation proceeds in three steps, as demonstrated by molecular dynamic simulations. First, upon application of the electric field, water defects appear in the cell membrane. If these defects are stable enough, they lead to the creation of a water file or hydrophobic pore through the membrane. Finally, the phospholipids in the vicinity this pore rearrange to yield a more stable and hydrophilic pore. However, the precise parameters that influence these three steps of the pore formation process are not identified yet, and therefore, the success yield of the technique remains low (<50%) (4).

Still, the technique of electroporation is widely used for various applications. This popularity is easily explained by the numerous advantages the technique presents compared to other permeabilization approaches. The technique itself is easy to setup, and does not require expensive equipment. Furthermore, electroporation is not toxic to the cells; it is highly reproducible and can be performed at a large scale, in a possibly automated way. Finally, this electrical approach is universal: it has a wide applicability in terms of cell lines that can be treated (including bacteria, plant and mammalian cells); any kind of foreign entities can be delivered into the cells (small molecules, drugs, plasmids, proteins, particles of different sizes...); and the technique can also be employed for extracting material out of the cells for analysis.

The main applications of electroporation are as follows: (1) gene therapy or cell vaccination (5), (2) cell engineering or modification to establish new cell lines for patch-clamp measurements or investigating intracellular signaling pathways, by delivery of either plasmids coding for exogenous proteins to be expressed by cells or siRNA to silence the expression of certain proteins (6, 7), (3) enhanced drug delivery e.g., for cancer treatment (also known as electrochemotherapy) (8), (4) particle delivery (9), (5) bacterial transformation or engineering of plant cells, and (6) protein delivery. For most of these applications, the electroporation treatment is carried out at the level of a whole cell population in bulk. Here, 1 mL of a cell suspension (ca. 10^6 cells) is placed in a cuvette equipped with two electrodes, on which a high voltage (kV range) is applied.

However, the bulk electroporation approach is not appropriate for all applications, and for some particular purposes, the treatment must be miniaturized to the single cell level (3, 10), or even to the subcellular level. Advantages found in this strategy are first the higher control on the electrical treatment to which the cell is exposed and the possible customization of the electrical parameters for each cell; this ultimately results in a higher success yield. Second, as the voltage is created across a single cell, the distance between the electrodes is much shorter (<100 μm), and 1–10 V voltages are typically used, eliminating the risks associated with the use of high voltages (kV range). In addition, this single cell approach is expected to yield basic knowledge on the processes underlying the mechanisms of pore formation. Furthermore, studying the response of single cells to the electroporation signal will generate information on the influence of numerous cell parameters (cell size, shape, membrane composition...) on the outcome of the treatment, and help elucidate the discrepancy found in cell populations. Finally, this single cell electroporation strategy can be seen as a novel noninvasive approach for single cell analysis to study signaling pathways; this relies on either sampling a small amount of the cell content (11, 12) or transfecting cells with a protein of interest coupled to a fluorescent reporter followed by tracking the protein position after cell exposure to given stimuli (13).

Two main approaches are reported in the literature for single cell electroporation: either using microelectrodes (14) (or micropipettes (15)) or in a microdevice (11–13, 16–26). In the first method, two microelectrodes (or micropipettes) are positioned in close vicinity to a cell with the help of micromanipulators. Subsequently, a voltage is applied between the electrodes to generate a well-defined electric field across the cell. This technique is labor-intensive, time-consuming and requires a skilled operator for precise positioning of the electrodes at a precise distance from the cell membrane. Still, one key advantage is that this approach is suitable for in situ treatment of adherent cells, in a natural environment. Alternatively, the electroporation protocol is implemented in a microfluidic device. The device typically includes micrometer-size structures not only to isolate a single cell from a population and trap it in a given location, but also to locally shape the electric field and create hot spots at the place where the cell is immobilized. This microfluidic-based strategy lends itself well to large-scale (e.g., by using an array of trapping sites) and automated cell poration, even if the protocol is performed at the single cell level (13, 19, 21). On other aspects, on-chip electroporation can also be coupled to other steps such as single cell analysis (e.g., using capillary electrophoresis), and single cell imaging to follow the postelectroporation fate of individual cells (27), which is important when cells have been transfected with genes (13).

In this chapter, we focus on this second approach where microfluidics is exploited for single cell electroporation, and only poration of mammalian cells is discussed, although some bacterial applications are included in Subheading 4.

Specifically, we describe a protocol for single cell electroporation using a glass-silicon microfluidic device that contains a series of independent poration sites (13). These sites consist of 4-μm wide slits where individual cells are trapped and successively exposed to an electric field created using integrated electrodes. The electroporation protocol is first optimized using various cell lines (K562, THP-1, and C2C12 cells) via the uptake or release of fluorescent dyes, and gene transfection is demonstrated using a plasmid coding for EGFP (Enhanced Green Fluorescent Protein). Finally, the single cell electroporation chip and protocol are applied to track the localization of a protein-kinase (ERK1) involved in intracellular signaling. For that purpose, cells are engineered to express an EGFP–ERK1 construct, and imaged on-chip using fluorescence microscopy to follow ERK1 activity upon cell stimulation. This work is an unprecedented example of an on-chip integrated protocol for cell transfection followed by cell imaging to elucidate signaling pathways, with all experiments being performed at the single cell level.

2. Materials

2.1. Cell Culture

1. Cells are cultured in conventional culture medium, and the nature of the medium depends on the cell type, as follows.
2. Human leukemia cells (THP1/K562) are cultured in RPMI-1640 medium supplemented with 10% heat-inactivated and filter-sterilized fetal calf serum, 100 IU/mL penicillin, 100 mg/mL streptomycin, 2 mM l-glutamine, and 250 mg/mL fungizone (RPMI+ medium).
3. C2C12 cells are cultured in DMEM medium supplemented with 10% fetal bovine serum (FBS), 100 IU/mL penicillin, 100 mg/mL streptomycin, 2 mM l-glutamine, 250 mg/mL fungizone, and 1% sodium pyruvate.
4. Mesenchymal stem cells (MSCs) are cultured in αMEM supplemented with 10% FBS, 1 ng/mL bFGF, 100 mg/mL penicillin, 100 IU/mL streptomycin, and 0.4 mmol/mL ascorbic acid.
5. All media, supplements and antibiotics are purchased from Invitrogen (Breda, The Netherlands), except for ascorbic acid which is purchased from Sigma (Zwijndrecht, The Netherlands).
6. A 0.25% (w/v) trypsin solution in PBS buffer is employed to harvest C2C12 and MSCs cells for culture and cell sample preparation before the electroporation.

2.2. Cell Staining

Cells (10^6/mL) are stained using a 1 μg/mL Calcein AM (Invitrogen, Breda, The Netherlands) solution in PBS buffer prepared by diluting 1,000 times a 1 mg/mL stock solution in DMSO. In this manner, electroporation can be detected via the leakage of the calcein out of the cells.

2.3. Electroporation Buffer

Electroporation experiments are carried out in a dedicated low-conductivity buffer composed of 10 mM HEPES, 140 mM NaCl, 2.68 mM KCl, 1.7 mM $MgCl_2$, and 25 mM glucose. The solution is maintained at a pH of 7.4.

2.4. DNA Solution

For gene transfection experiments, DNA is loaded in the microfluidic chip after trapping of the cells in the device. The working solution introduced in the chip has a concentration of 100 ng/mL; it is prepared by diluting a MilliQ water-based DNA stock solution (450 ng/μL) in electroporation buffer. Two plasmids are employed here: one plasmid coding for the enhanced Green Fluorescent Protein (EGFP) and one plasmid coding for a protein construct composed of ERK1 (signaling protein) and a fluorescent reporter (EGFP).

2.5. Microchip

The essential part of single cell electroporation in a microfluidic format is the microdevice itself. This device must include a number

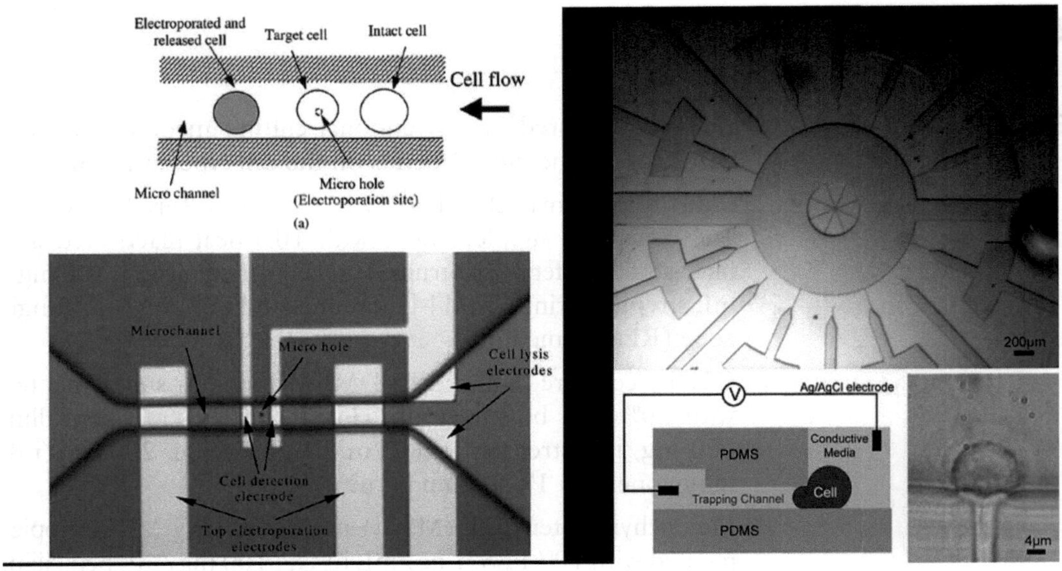

Fig. 2. Examples of miniaturized and microfluidic devices for single cell electroporation. *Left*: the device consists of a main channel where cells flow and at the bottom of which a microhole is added; when a cell passes by, it is trapped in the microhole and exposed to an electrical treatment applied using two integrated electrodes. Subsequently, the cell is released and the same treatment is applied to the next cell flowing in the channel (Reprinted from (Huang, 2003) with permission from Elsevier). Reproduced by permission of The Royal Society of Chemistry. *Right*: the device includes one central chamber connected to two wide channels for loading and removal of the cell solution, as well as an array of channels radiating from the chamber and connected to it via low μm^2-cross-section channels, used for cell trapping (*top*). External electrodes are inserted in the main channel and side channels of the devices. *Bottom*: schematic representation of cell trapping in the constriction channels, and enlarged view of a trapped cell. Reprinted from (Khine, 2005). Reproduced by permission of The Royal Society of Chemistry.

of specific structures, as follows. First, it contains a microfluidic channel where the cell solution is introduced. Second, this channel presents a single trapping site or an array of those, disposed in a planar way (at the bottom of the channel) or in a lateral fashion (along one wall of the channel) in which cells are isolated. For actual trapping of cells, these structures are connected on their backside to another microfluidic structure (e.g., channel or chamber) from which a negative pressure is applied. Finally, electrodes are necessary to create a localized high electric field across the cell(s). They can either be integrated in the device or externally introduced in the reservoirs.

Several single cell electroporation devices are reported in the literature, as illustrated in Fig. 2, with one or several trapping sites, lateral or bottom trapping of cells, and integrated or external electrodes. Here, we focus on one particular device developed in our group.

This single cell electroporation chip consists of two microfluidic channels etched in silicon and connected with each other through an array of nine trapping structures (20 μm width; 4 μm depth),

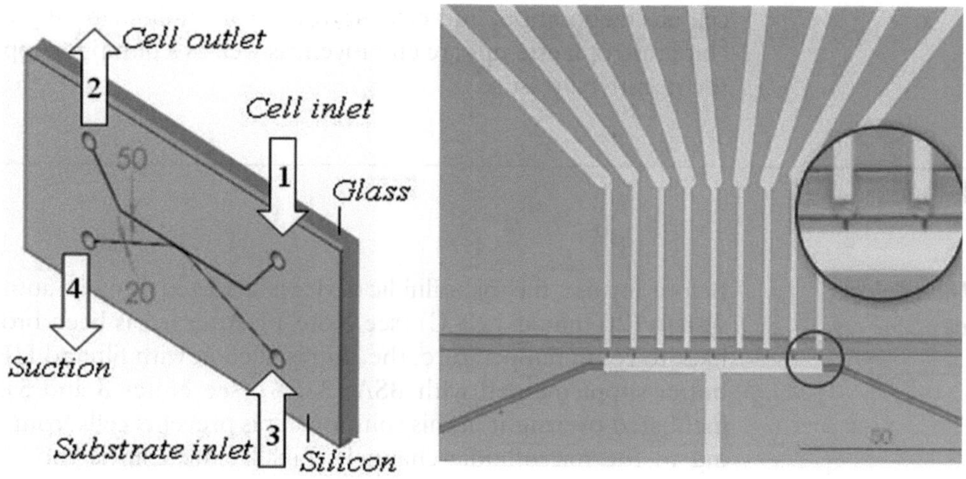

Fig. 3. Single cell electroporation device employed in this chapter including an array of nine independent trapping sites. (**left**) Cartoon of the microfluidic chip showing the two microfluidic channels for cell flow and monitoring cell trapping separated by an array of trapping sites. (**right**) Top view of the microfluidic chip showing the two channels and the electrodes to individually address the nine trapping sites where single cells are immobilized; *inset*: enlarged view of two trapped cells ((Valero, 2008) Reproduced by permission of The Royal Society of Chemistry). Enlarged view on a few trapping sites located between the two microfluidic channels (SEM picture).

shown in Fig. 3. One (wide) channel (50 μm width; 15 μm depth) is used for the introduction of cells and the various solutions in the chip, while the other channel (20 μm width; 15 μm depth) is employed to suck the cells in the traps. This silicon fluidic substrate is bonded to a Pyrex substrate on which Pt electrodes are sputtered: a main and common electrode for all sites as well as separate electrodes for individual addressing of the nine trapping sites. This device is produced by microfabrication techniques in a dedicated clean-room environment; more details on the fabrication process can be found elsewhere (13).

2.6. Experimental Setup

For electroporation experiments, the microfluidic chip is housed in a dedicated chip-holder, which includes fluidic access to the reservoirs of the chip and integrated electrical connections. The chip-holder is connected via an in-house designed interface to a computer equipped with function generator (NI 5041 National Instruments) and acquisition cards (NI PCI-6221, National Instruments) for application of the electroporation signals on the nine independent electrodes and recording of electrical signals. Electroporation experiments are monitored using a LabView interface that controls both NI cards. The chip-holder is placed on the stage of a microscope equipped with an epifluorescence unit and a controlled CCD camera for optical monitoring of the experiments (cell trapping, cell electroporation, and cell survival). For fluidic handling in the chip, micropipette tips are employed, as well as a pump for applying the negative pressure.

3. Methods

3.1. Microfluidic Device

Before any use, the microfluidic device is sterilized using an autoclave system (20 min at 122°C) (see Note 1). After it has been brought back to room temperature, the chip is filled in with filtered HEPES buffer supplemented with BSA (3–5%) (see Notes 3 and 5), and incubated overnight in this solution. This prevents cells from sticking in the microfluidic channels. Finally, the chip is thoroughly rinsed with filtered HEPES buffer.

3.2. Cell Sample Culture and Preparation

Cells are cultured in the aforementioned media, and medium is refreshed every 3–4 days. C2C12 and MSC cells are first washed with PBS buffer to remove dead cells as well as any protein present in the culture flask. Subsequently, the cells are incubated with a trypsin solution (0.25% w/v) for a few minutes, and complete growth medium is subsequently added to inactivate the trypsin. Following this, the resulting cell solution is centrifuged and cells are resuspended in fresh and prewarmed medium in case of culture, or in electroporation buffer in case of experimentation. For THP-1 and K562 cells, the procedure is the same, but without any trypsin treatment as those cells are already in suspension.

THP-1, K562, and C2C12 cells have been employed to optimize the study of the electroporation process using fluorescence assays (calcein release, PI entry), and C2C12 cells and MSCs for gene transfection experiments.

For some experiments, cell electroporation is detected optically through the release of a dye out of the cells. This dye (typically Calcein AM) is previously loaded in the cells by incubating them in a 1 μg/mL solution for 30–60 min. Thereafter, the cells are washed thoroughly twice with PBS solution at room temperature, and finally resuspended in the electroporation buffer.

3.3. Cell Loading

All reservoirs are filled with electroporation buffer (100 μL), and in reservoir 1, 100 μL of the cell suspension (10× diluted) in electroporation buffer is placed. By removing the buffer (ca. 50 μL) in reservoir 2, a cell flow is created in the main channel. Subsequently, a mild suction (1–2 psi) is applied from reservoir 3 or 4 to attract the cells in the trapping structures. If more than one cell is retained in the trapping structures, a higher flow is created in the main channel to remove the excess of cells, while maintaining the negative pressure in the traps. Once all traps are filled with a single cell, the negative pressure is switched off.

3.4. Preelectroporation Preparation Steps

After trapping of the cells, the buffer in the main channel is refreshed to introduce new buffer supplemented with either plasmids (for gene transfection experiments) or a membrane exclusion dye (for optical detection of pore formation based on the entry of a fluorescent dye in the cells). In the former case, cells are incubated for ca. 10 min with the DNA solution before application of the electroporation signal (see Note 24). This incubation time promotes the formation of DNA–cell adducts, and ultimately enhances the entry of DNA in the cell upon application of the electric field.

3.5. Cell Electroporation

The electroporation signal is applied on the cells trapped in the device. As the electrodes are independently addressable, the cells are treated individually and successively. Typically, an electroporation signal consists of one pulse of 6-ms duration and amplitudes starting from 1 V (see Notes 12–15). If no electroporation is detected (no release of calcein or no uptake of PI, see Subheading 3.7), a second electroporation signal is applied with a longer pulse duration (+1 ms) or a higher amplitude (+0.5 V) until pore formation is observed optically. For C2C12 and MSCs cells, an optimal electroporation protocol consists of one pulse of 2 V and 6 ms, whereas other cell lines require higher field strength before uptake of dye is detected, and this still depends on the type of the cell lines, as shown in Fig. 4.

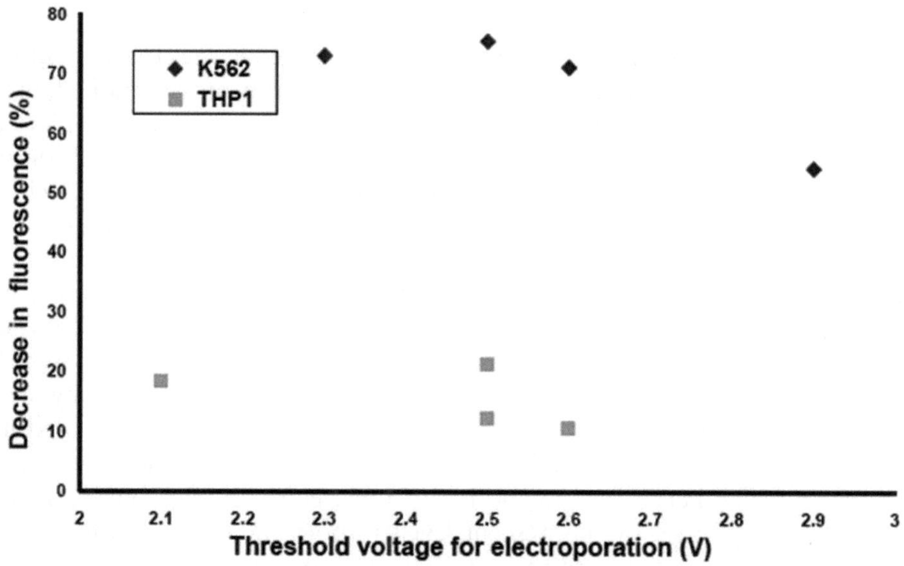

Fig. 4. Discrepancy in the required electrical parameters required to electroporate cells from two different cell lines (K562 and THP1) measured through the release of calcein out of the cells after application of an electrical treatment and the value at which cell poration is detected. K562 cells are more porated than THP-1 cells for a similar electrical treatment (6 pulses of 1 ms, 2–2.9 V amplitude), as observed by the lower decrease in fluorescence intensity measured in the cells (20% for THP1 cells against >70% for K562 cells). This discrepancy can be explained by a difference either in cell size (16 μm diameter for K562 cells against 12 μm for THP 1 cells) or in cell membrane properties (Courtesy Ms. V. Stimberg).

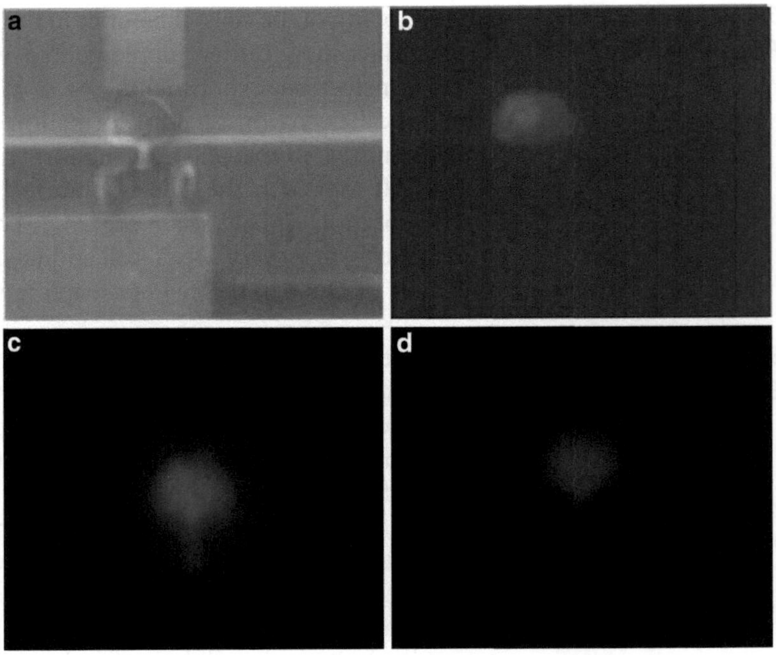

Fig. 5. Examples of fluorescence-based monitoring of cell poration using either the entry of PI into C2C12 cells (*top*), or the release of calcein out of THP-1 cells (*bottom*). C2C12 cells are porated using a single pulse (6 ms, 2 V) ((Valero, 2008) Reproduced by permission of The Royal Society of Chemistry) while THP-1 cells are treated using a series of six pulses (3 ms, 2.4 V) (Courtesy Ms. V. Stimberg). Pictures (**a**) and (**c**) show the cells before the treatment, and pictures (**b**) and (**d**) after application of the electrical pulse(s).

3.6. Postelectroporation Process

For fluorescence assays, cells are imaged in the electroporation buffer (see Subheading 3.7), and once experiments are finished, the chip is cleaned and washed thoroughly to be used again (see Note 2). For DNA transfection experiments, the cells are incubated for an additional 10 min to complete entry of the plasmids in the cells, before the solution is changed to culture medium (see Note 24).

3.7. Detection of Electroporation

Fluorescent membrane integrity markers are mostly used to detect pore formation in the cell membrane (see Note 20). These dyes do not usually cross the membrane unless the latter is damaged. Fluorescence-based detection is done twofold; either by measuring the release of a dye (e.g., calcein AM) previously loaded into the cells (21), or by detecting the entry of another dye (e.g., PI or YOYO-1) (13) added in the electroporation buffer. In the first case, the fluorescent levels in both the cells and the surrounding medium are measured after exposure of the cells to the electric field (see Note 20). In the second case, the dye is a DNA intercalating agent, and upon entry in cells, it moves to the nucleus and binds to DNA to give a bright fluorescent signal. Figure 5 illustrates this principle of fluorescence-based monitoring of cell electroporation, using both approaches of dye loading in the cell (Fig. 5a–b) and dye release out of the cells (Fig. 5c–d). However, all dyes (calcein, PI, and YOYO-1) are markers for cell death, so a better option would be to use a sequential combination of two dyes (calcein and PI) where the second dye would be added in the medium 30 min after electroporation (when pores are expected to be closed). Thereby, it would be possible to distinguish between viable cell poration (calcein release, no PI uptake) and cell death (calcein release, PI uptake). Alternatively, small ions and FITC-conjugated Dextran particles can be employed to probe the size of the pores created upon application of the electrical signal (see Note 21).

3.8. Gene Transfection

Other fluorescence assays are based on the transfection of a gene coding for a fluorescent protein such as GFP; this enables to demonstrate to only that cells are porated, but also that they are functioning properly as they are able to produce proteins from a gene. For gene transfection experiments, all solutions used (cell solution, electroporation buffer, DNA solution) are cooled on ice before being introduced in the chip (see Note 4). After the poration process and once warm medium has been introduced to replace the electroporation buffer, the chip is removed from its holder, and placed in a Petri dish, covered with warm medium. The Petri dish is kept under controlled conditions (37°C, 5% CO_2) in the incubator for one or several days. Typically, after 24 h, cells are imaged to check for protein expression (positive green fluorescent signal) (Fig. 6), and the fluorescence level in the cell is quantified to determine the amount of plasmids loaded in the cell (13).

3.9. Application: Studying the ERK1 Signaling Pathway

As mentioned in the introduction, one promising application of single cell electroporation in a microfluidic platform is the elucidation of signaling pathways. For that purpose, a plasmid coding for a protein construct is loaded in MSCs cells; here, a construct composed of a kinase protein (ERK1) and EGFP is employed, where EGFP acts as a fluorescent reporter for the localization of the kinase inside the cell (13). As before, one day after cell poration, cells are imaged to check for the expression of the construct protein, and a uniform staining is observed in the whole cell.

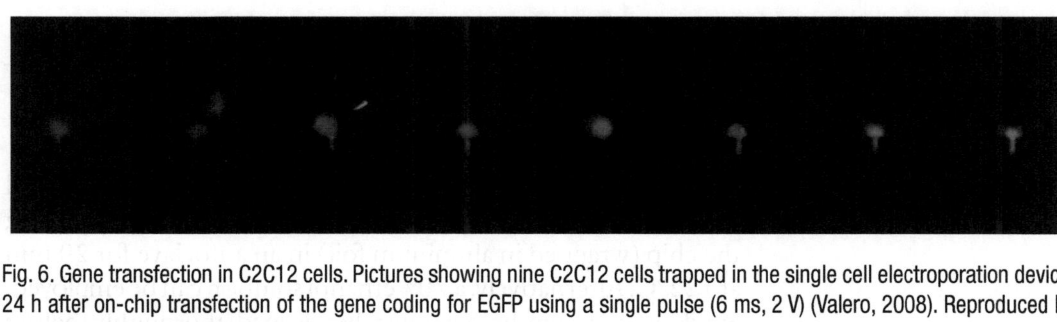

Fig. 6. Gene transfection in C2C12 cells. Pictures showing nine C2C12 cells trapped in the single cell electroporation device, 24 h after on-chip transfection of the gene coding for EGFP using a single pulse (6 ms, 2 V) (Valero, 2008). Reproduced by permission of The Royal Society of Chemistry.

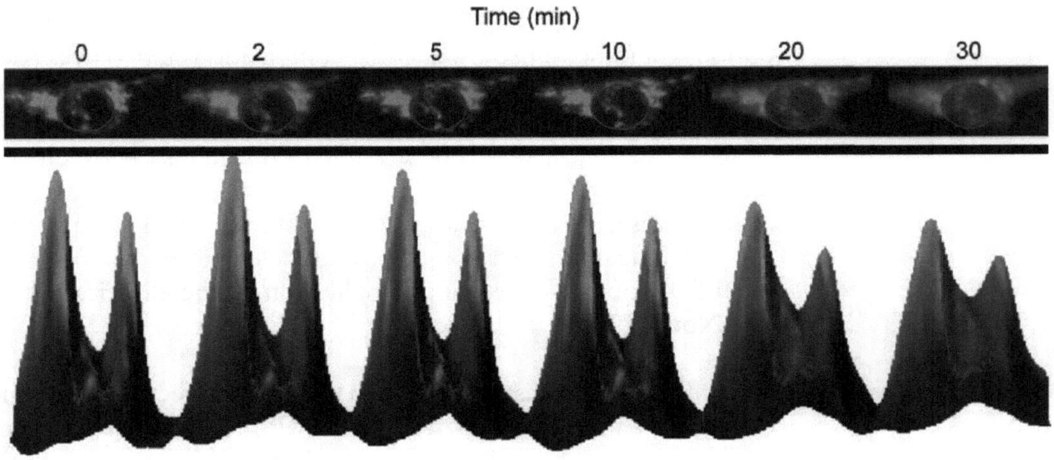

Fig. 7. Imaging the ERK1 signaling pathway using single cell electroporation technology. After having been transfected with a gene coding for the ERK1–EGFP construct (EGFP playing the role of a fluorescent reporter for the protein kinase ERK1), a cell is placed under starvation conditions until ERK1 is solely localized in the cytoplasm (time 0). Subsequently, the cell is stimulated via exposition to a growth factor (FGF-2) for activation of ERK1 and its translocation to the nucleus; this is visualized using real-time imaging by a progressive shift of the fluorescence from the cytoplasm to the nucleus. *Top*: Fluorescent images of a single cell after exposure to FGF-2; *Bottom*: 3D representation of the fluorescent intensity in the cell; as time passes the fluorescent level decreases in the cytoplasm and increases in the nucleus (Valero, 2008). Reproduced by permission of The Royal Society of Chemistry.

Using this protein construct, the ERK1 signaling pathway initiated by the binding of growth factors (FGF-2) to the cell membrane is studied. This binding causes ERK1 activation and its translocation to the nucleus to trigger gene expression (28). In a first step, MSCs cells are kept under "starvation" conditions, in a serum-poor medium (1% instead of 10%) for one day, resulting in the full inactivation of ERK1: after 24 h, it is solely located in the cell cytoplasm, as seen by the green fluorescent pattern limited to the cell cytoplasm. Thereafter, the solution in the chip is changed for a medium containing FGF-2 (10 ng/mL). To follow the activation of ERK1, the cells are imaged continuously for 30 min; Fig. 7 shows the progressive translocation of ERK1 to the nucleus upon FGF-2-based activation, visualized by a shift of the green fluorescent signal from the cytoplasm to the cell nucleus.

4. Notes

General Matters

1. Sterilization of the microchip before any experiment is essential for the outcome of the experimentation. This is done by placing the chip (wrapped in aluminum foil) in an autoclave for 20 min at 122°C. Alternatively, a 70% ethanol solution can be employed. Similarly, the chip-holder must be cleaned thoroughly before and after use with a 70% ethanol solution, followed by rinsing in MilliQ water.

2. Before the chips are reused, they must be thoroughly cleaned to remove any cell debris, by incubating them in a trypsin solution (0.25% w/v) overnight in the incubator. Thereafter, the chip is sterilized anew in the incubator.
3. To avoid clogging issue, all solutions are filtered before being introduced in the chips (0.2-μm nonprotein binding filters, Millipore).
4. A low experimental temperature (0–4°C) is preferred to delay pore resealing and subsequently enhance cell loading efficiency. This is especially important for transfection of genes that penetrate slowly into the cells. This is achieved by placing the chip in a cold environment and by chilling all solutions before use.

Solutions

5. The cell suspension can be supplemented with BSA (3–5%) to prevent cells from aggregating with each other and being trapped as clusters.
6. The pH of the electroporation buffer must be chosen as close as possible to the intracellular pH (e.g., pH 7.2). Similarly, the buffer composition should mimic the composition of the cytoplasm to avoid extensive cell swelling (that would easily lead to cell death).
7. The ionic strength of the buffer should be as low as possible. On the one hand, the ionic strength determines the resistivity of the solution, and subsequently the time constant (RC) of the electroporation process (29). On the other hand, a high ionic content would favor arcing phenomena at the electrodes, which can affect cell viability. The presence of small ions such as Ca^{2+} and Mg^{2+} also promote cell recovery after the electroporation treatment (29).
8. The DNA solution concentration must be adjusted to enhance cell transfection (29).

Electrodes

9. If electrodes are integrated in the chip, the biocompatibility of the materials used to fabricate them must be known. Electrodes are prone to release ions and particles which may be toxic to cells (30).
10. The position of the electrodes must be wisely chosen. When the pulse amplitude is too high, a too small interelectrode distance often leads to bubble formation (through the electrolysis of water), a change in the pH and the production of chemicals (e.g., ROS) which are toxic to cells (3). Subsequently, the addition of a constricted area appears as a more judicious approach to locally create a high electric field.
11. Electrodes suffer from corrosion, especially in case of electrochemical reactions. For repeated use, they can be coated with an insulating layer such as Teflon (31). However, this coating influences the electric field distribution.

Electrical Parameters

12. Mostly, square DC pulses (ns to μs) or exponentially decay pulse (ms) are employed for cell poration. Alternatively, using AC signals, risks for water electrolysis are decreased (32), the signal amplitude is lowered (33) and higher cell transformation yields have been reported (29).
13. The pulse parameters must be optimized depending on the application. Drug delivery requires short and high pulses while DNA transfection is enhanced using lower and shorter pulses, which promote DNA penetration in cells (34). Furthermore, when larger molecules or particles have to be injected in the cells, the pulse amplitude must be increased to create larger pores (35).
14. Similarly, the pulse parameters must be adjusted as a function of the cell type and a number of cell parameters (shape, size, and membrane composition) (36). This is particularly relevant for bacteria and plant cells which possess a different shape and membrane properties than mammalian cells. However, if the cells are truly trapped in a microhole, their size and shape should have little influence on the electroporation outcome.
15. The distribution of the electric field in the device must be carefully studied and modeled to determine the electric field strength across the cells.

Trapping Structures and Trapping Protocol

16. The electrical signal required for cell electroporation correlates with the size of the trapping site; the smaller the site, the higher the electric field across the trap, and so the lower the voltage applied. As already mentioned, one can benefit from this focusing effect when the interelectrode spacing is increased to alleviate unwanted electrochemical reactions.
17. The size of the trapping structures must be adapted for single cell trapping. A too large site leads to multiple cell trapping, and this affects the electroporation treatment. Conversely, cells are not tightly trapped in too small holes, and they can be easily released. Consequently, a single cell electroporation approach is difficult to be implemented for bacteria which have a size of a few microns, as holes in the submicrometer range are required to trap them.
18. Cell squeezing through the trap is observed when the trapping site is too large, or if the suction pressure is too high.
19. The trap size and the suction protocol determine the sealing quality of the cell in the trapping site. In case of a bad sealing, a leakage pathway exists so that a higher poration signal must be applied. Conversely, with a good sealing, the electrical treatment is milder. Furthermore, in this latter case, the cell poration voltage does no longer depend on the cell size and shape; only their membrane composition and their "fragility" influence the poration process.

Monitoring of Pore Formation: Fluorescent or Electrical Approach?

20. Pore formation is mostly monitored using fluorescence microscopy techniques through the release/entry of dyes out of/into the cells. However, in the former case, care should be taken to calibrate the photobleaching behavior of the dyes (negative control) and to also determine the fluorescence level in the solution in the close vicinity of the cells before concluding on cell poration.
21. Fluorescence-based assays can also rely on other probes. For instance, Dextran particles coupled to a fluorophore (FITC) which are available with different sizes (or molecular weights) are employed to probe the size of the pores created in the membrane as a function of the strength of the poration signal (35).
22. In general, fluorescence-based assays are invasive as they imply loading of the cells with fluorescent probes (before or upon cell poration). Alternatively, electrical measurements (patch-clamp recording) are performed to detect pore formation, which results in changes in the cell impedance (37); the latter employs the same electrodes for the electroporation and measurements. This electrical approach is particularly interesting to automate gene transfection processes through real-time monitoring of the electroporation process (20). Furthermore, using this approach, pores which are too small to enable the transport of dye molecules through the cell membrane can be detected.
23. However, for electrical-based detection of pore formation, a good sealing of the cell in the trapping structure ("giga-ohm seal") is mandatory as the presence of a leakage current precludes the detection of pore formation (21).

Gene Transfection Application

24. As stated in the methods, it is essential to incubate on chip the cells in the DNA solution both before and after application of the electroporation signal. In the first case, DNA comes in close vicinity to the cell and has a greater chance to be pulled into the cells upon application of the electrical signal. The role of the second incubation time is to give enough time to the DNA to fully penetrate into the cells.
25. Alternatively, DNA entry in the cells can be promoted with the application of a second lower and longer DC voltage (38), establishing thereby an electroosmotic flow which pulls the negatively charged plasmids into the cells.
26. For some cells, the transfection yield is limited as plasmids injected into the cells by electroporation do not reach the nucleus. In this case, a second series of shorter pulses (ns-range) can be applied to porate the membrane of the nucleus and enhance DNA delivery to the nucleus (39).

Alternative Single Cell Electroporation Approaches

27. The single cell electroporation approach described in this chapter is limited in terms of volume of cells transfected. The current device contains nine trapping sites, and more generally, this approach is limited by the number of electrodes that can be integrated in a single device. Therefore, two other approaches are preferred for treatment of larger amounts of cells, either using a single microhole, where cells are sequentially trapped, porated, and released (19), or using a flow-through device that contains a short constriction through which cells flow as a single-cell line, while a DC voltage is applied in the whole channel (23). However, these two approaches are not compatible with postelectroporation single cell tracking.
28. As mentioned in the introduction, single cell electroporation has also been demonstrated using two microelectrodes positioned close to the cell (14). In this case, one of the electrodes can be replaced by a pipette filled with the solution to be delivered in the cell for localized and subcellular treatment of cells (15).
29. Subcellular treatment has been achieved also using a nanoporous membrane on which cells are grown (40). The electric field is focused through the nanopores, possibly on specific areas in the cell membrane. Furthermore and interestingly, this protocol is applicable to adherent cells directly in their natural environment. Here, cells are grown as a monolayer on the large electrode and a counter-electrode (micropipette, microelectrode) is brought in close contact with the cells from the solution side for localized poration of adherent cells. However, this strategy resembles the microelectrode-based approach and is limited in terms of throughput.

Acknowledgment

The Single Cell Electroporation Chip presented in this chapter has been developed in the frame of the NanoScan project supported by the Dutch Technology Foundation STW. The authors would like to thank Dr. A. Valero, Dr. I. van Uitert, and Ms. V. Stimberg who have worked on this project, J. van Nieuwkasteele and P. ter Braak for technical assistance, and V. Stimberg and Dr. J.-P. Frimat for proofreading.

References

1. Neumann, E. and Rosenhec, K. (1972) Permeability changes induced by electric impulses in vesicular membranes. *J Membr Biol* 10, 279–290.
2. Zimmermn, U., Pilwat, G., and Riemann F. (1974) Dielectric-breakdown of cell-membranes. *Biophys J* 14, 881–899.
3. Wang, M.Y. et al. (2010) Single-cell electroporation. *Anal Bioanal Chem* 397, 3235–3248.
4. Lee, W.G., Demirci, U., and Khademhosseini, A. (2009) Microscale electroporation: challenges and perspectives for clinical applications. *Integr Biol* 1, 242–251.
5. Bloquel, C. et al. (2004) Plasmid DNA electrotransfer for intracellular and secreted proteins expression: new methodological developments and applications. *J Gene Med* 6, S11–S23.
6. Aberg, M.A.I. et al. (2001) Selective introduction of antisense oligonucleotides into single adult CNS progenitor cells using electroporation demonstrates the requirement of STAT3 activation for CNTF-induced gliogenesis. *Mol Cell Neurosci* 17, 426–443.
7. Golzio, M. et al. (2005) Inhibition of gene expression in mice muscle by in vivo electrically mediated siRNA delivery. *Gene Therapy* 12, 246–251.
8. Belehradek, M. et al. (1993) Electrochemotherapy, a new antitumor treatment – 1st clinical phase-i-ii trial. *Cancer* 72, 3694–3700.
9. Yoo J.S., *et al.* (2010) In vivo imaging of cancer cells with electroporation of quantum dots and multispectral imaging. *J. Appl. Phys.* 107 (12).
10. Olofsson, J. et al. (2003) Single-cell electroporation. *Curr Opin Biotech* 14, 29–34.
11. Bao, N., Wang, J., and Lu, C. (2008) Microfluidic electroporation for selective release of intracellular molecules at the single-cell level. *Electrophoresis* 29, 2939–2944.
12. Wang, J. et al. (2008) Detection of kinase translocation using microfluidic electroporative flow cytometry. *Anal Chem* 80, 1087–1093.
13. Valero, A. et al. (2008) Gene transfer and protein dynamics in stem cells using single cell electroporation in a microfluidic device. *Lab Chip* 8, 62–67.
14. Lundqvist, J.A. (1998) Altering the biochemical state of individual cultured cells and organelles with ultramicroelectrodes. *Proc Natl Acad Sci USA* 95, 10356–10360.
15. Wang, M.Y., Orwar, O., and Weber, S.G. (2009) Single-Cell Transfection by Electroporation Using an Electrolyte/Plasmid-Filled Capillary. *Anal Chem* 81, 4060–4067.
16. Bao, N., Zhan, Y.H., and Lu, C. (2008) Microfluidic electroporative flow cytometry for studying single-cell biomechanics. *Anal Chem* 80, 7714–7719.
17. Fox, M.B. et al. (2006) Electroporation of cells in microfluidic devices: a review. *Anal Bioanal Chem* 385, 474–485.
18. Huang, Y. and Rubinsky, B. (2001) Microfabricated electroporation chip for single cell membrane permeabilization. *Sens Act A* 89, 242–249.
19. Huang, Y. and Rubinsky, B. (2003) Flow-through micro-electroporation chip for high efficiency single-cell genetic manipulation. *Sens Act A* 104, 205–212.
20. Khine, M. et al. (2007) Single-cell electroporation arrays with real-time monitoring and feedback control. *Lab Chip* 7, 457–462.
21. Khine, M. et al. (2005) A single cell electroporation chip. *Lab Chip* 5, 38–43.
22. Kurosawa, O. et al. (2006) Electroporation through a micro-fabricated orifice and its application to the measurement of cell response to external stimuli. *Meas Sci Technol* 17, 3127–3133.
23. Wang, H.Y. and Lu, C. (2008) Microfluidic electroporation for delivery of small molecules and genes into cells using a common DC power supply. *Biotech Bioeng* 100, 579–586.
24. Xiao, K. et al. (2010) Electroporation of microdroplet encapsulated HeLa cells in oil phase. *Electrophoresis* 31, 3175–3180.
25. Zhan, Y.H. et al. (2009) Electroporation of Cells in Microfluidic Droplets. *Anal Chem* 81, 2027–2031.
26. Zhu, T. et al. (2010) Electroporation based on hydrodynamic focusing of microfluidics with low dc voltage. *Biomed Microdevices* 12, 35–40.
27. Sims, C.E. et al. (1998) Laser-micropipet combination for single cell analysis. *Anal Chem* 70, 4570–4577.
28. Brunet, A. et al. (1999) Nuclear translocation of p42/p44 mitogen-activated protein kinase is required for growth factor-induced gene expression and cell cycle entry. *EMBO J* 18, 664–674.
29. Prasanna, G.L. and Panda, T. (1997) Electroporation: Basic principles, practical considerations and applications in molecular biology. *Bioproc Eng* 16, 261–264.
30. Loomishusselbee, J.W. et al. (1991) Electroporation can cause artifacts due to solubilization of cations from the electrode plates – aluminum ions enhance conversion of inositol 1,3,4,5-tetrakisphosphate into inositol 1,4,5-trisphosphate in electroporated l1210 cells. *Biochem J* 277, 883–885.

31. Lee, S.W. and Tai, Y.C. (1999) A micro cell lysis device. *Sens Act A* 73, 74–79.
32. Lu, H., Schmidt, M.A., and Jensen, K.F. (2005) A microfluidic electroporation device for cell lysis. *Lab Chip* 5, 23–29.
33. Xie, T.D., Sun, L., and Tsong, T.Y. (1990) Study of mechanisms of electric field-induced DNA transfection. 1. DNA entry by surface binding and diffusion through membrane pores. *Biophys J* 58, 3–19.
34. Rols, M.P. (2006) Electropermeabilization, a physical method for the delivery of therapeutic molecules into cells. *Biochim Biophys Acta – Biomembranes* 1758, 423–428.
35. He, H.Q., Chang, D.C., and Lee, Y.K. (2007) Using a micro electroporation chip to determine the optimal physical parameters in the uptake of biomolecules in HeLa cells. *Bioelectrochem* 70, 363–368.
36. Agarwal, A. et al. (2007) Effect of cell size and shape on single-cell electroporation. *Anal Chem* 79, 3589–3596.
37. Ryttsen, F. et al. (2000) Characterization of single-cell electroporation by using patch-clamp and fluorescence microscopy. *Biophys J* 79, 1993–2001.
38. Ionescu-Zanetti, C., Blatz, A., and Khine, M. (2008) Electrophoresis-assisted single-cell electroporation for efficient intracellular delivery. *Biomed Microdevices* 10,113-116.
39. Schoenbach, K.H., Beebe, S.J., and Buescher, E.S. (2001) Intracellular effect of ultrashort electrical pulses. *Bioelectromagnetics* 22, 440–448.
40. Ishibashi, T. et al. (2007) A porous membrane-based culture substrate for localized in situ electroporation of adherent mammalian cells. *Sens Act B* 128, 5–11.

Chapter 8

Perfusion Culture of Mammalian Cells in a Microfluidic Channel with a Built-In Pillar Array

Chi Zhang

Abstract

In vitro culture of mammalian cells is fundamental to various biological studies such as single cell analysis, pathological research, and drug/therapy development. However, there are limitations with the current in vitro cell culture methods. Cells tend to lose their specific functions due to the lack of a cellular microenvironment when they are maintained under standard culture conditions. Microscale devices can be a novel tool to reestablish a cellular microenvironment for culturing mammalian cells in vitro and maintaining their differentiated functions. Different microscale cell culture techniques have been developed to suit different biological applications. This chapter describes a method to trap and culture mammalian cells in a microfluidic channel with a built-in pillar array.

Key words: Microfluidics, 3D Cell culture, Microenvironment, Cell analysis

1. Introduction

Cell culture is fundamental to both basic biological research and biotechnology development (1, 2). Majority of our current understandings of the cellular structures and functions are based on cell cultures (3, 4). The conditions and microenvironments in which the cells are cultured greatly influence their behaviors (5). In the cellular microenvironment, cells form strong interactions with their neighboring cells (6) and are primarily protected by the extracellular matrix (ECM), which contains insoluble proteins (e.g., collagen, fibronectin, and laminin) (7) and provides mechanical support and stimuli to the cells (5, 8). In the microenvironment, different kinds of soluble factors are presented in the cell vicinity within a gradient, which is the major source of chemical signals to the cells (9, 10). Creating a controllable microenvironment for in vitro cultured cells is critical to their functional enhancement and thus the downstream applications.

Sara Lindström and Helene Andersson-Svahn (eds.), *Single-Cell Analysis: Methods and Protocols*, Methods in Molecular Biology, vol. 853, DOI 10.1007/978-1-61779-567-1_8, © Springer Science+Business Media, LLC 2012

Microscale cell culture techniques have been developed to allow more precise control of the extracellular microenvironment. By making use of microscale experimental tools, cell–cell, cell–matrix, and cell–soluble factor interactions can be defined and the influences mechanical forces have on cells can be reestablished. Static microscale cell culture (e.g., micropatterning technology) has been successful in enabling novel experimentations that control cell–cell and cell–matrix interactions (11, 12). Likewise, microfluidic perfusion culture allows novel experimentations for controlling the microenvironment by which cellular phenotypes are affected (13). Laminar flow in microfluidic systems enables the delivery of multiple streams containing different soluble molecules at the cellular or subcellular levels in a more controlled way (14). Meanwhile, shear stress can be applied via laminar flow to cells as mechanical stimuli (15). For tissue types that are highly perfused in vivo, such as liver and kidney, cells are in close proximity with the microvascular network. A more physiologically relevant in vitro model can better preserve the in vivo-like phenotypes of cells from these tissues and therefore enables the acquisition of biologically meaningful data from cell-based assays (16, 17). Microfluidic perfusion culture may facilitate establishing a defined, artificial microenvironment for these highly perfused cells by continuously controlling the supply and removal of soluble factors (18). Such defined environments are unobtainable in static culture, where the background of soluble factors and nutrients changes constantly over time. For higher-throughput analysis, fluid handling operations can readily be implemented by integrating microfluidic valves, mixers, or gradient generators (19).

This chapter describes a method of trapping and culturing mammalian cells in a microfluidic channel with a build-in pillar array. The microfluidic channel is fabricated using photolithography and replica molding. By using this microscale cell culture technique, a biomimetic microenvironment can be created. The microfluidic channel can further be used as a generic cell culture platform for a variety of cell types to perform further biological investigations such as single cell manipulation and analysis, drug testing, etc.

2. Materials

Prepare all solutions at room temperature using deionized water and reagents at the analytical grade. Store all prepared solutions at 4°C (unless indicated otherwise). Diligently follow all waste disposal regulations when disposing waste materials.

2.1. Microfabrication on Silicon Wafers

1. Design the microfluidic channel with the pillar array using AutoCAD (Autodesk, USA). The microfluidic channel is 1 cm (length) × 600 μm (width) × 100 μm (height) with two inlets and one outlet. An array of 30 μm × 50 μm elliptical micropillars with 20 μm gap size locates at the center of the microfluidic channel, creating a cell culture compartment and two flanking channels (Fig. 1).
2. Print the design onto a mask. Coat photoresist onto the silicon wafer (4″) and bake the coated wafer to let the photoresist be cured (see Note 1).
3. Insert both the mask and the baked wafer into a mask aligner (EV Group, USA) and the mask should be positioned over the wafer. Shine ultraviolet light vertically from above the mask. The short-waved ultraviolet light will go through the clear portions of the mask transferring the pattern onto the photoresist (see Note 2).
4. Immerse the wafer into developer solutions so that unwanted photoresist is washed away and the pattern of the microfluidic channel is obtained on the wafer (see Note 3).
5. Etch away the area on the wafer that is not protected by photoresist through standard deep reactive ion etching (DRIE) process (Oxford Instruments Plc, UK). The etching takes 20 min.

2.2. Medium, Buffers, Solutions and Equipment for Cell Culture

1. Culture media: 900 mL of low glucose basal Dulbecco's modified Eagle medium (DMEM) without phenol red (Gibco Invitrogen, USA), 100 mL of sterile fetal bovine serum (FBS) (Gibco Invitrogen, USA), 100 μg of penicillin/streptomycin (Sigma, USA), 1.5 g of sodium bicarbonate (Merck, Germany). Under sterile conditions, mix the components mentioned above and store (see Note 4).
2. 1× Phosphate buffered saline (PBS): 800 mL of water, 8 g of NaCl, 0.2 g of KCl, 1.44 g of Na_2HPO_4, 0.24 g of KH_2PO_4. Mix all the components at room temperature; adjust pH to 7.3–7.4. Fill the solution to 1 L with water and sterilize by autoclaving.

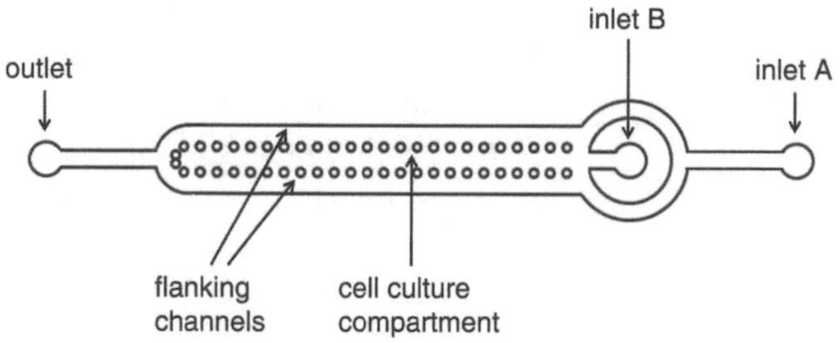

Fig. 1. Schematic illustration of the microfluidic channel. The micropillar array separates two flanking channels from the cell culture compartment at the center.

3. Trypsin solution: 0.25% Trypsin–EDTA in 1× PBS, sterilize by filtering.
4. Collagen type I solution: 3.0 mg/mL of collagen type I (Sigma, USA) in 1× PBS.
5. Bovine serum albumin (BSA) solution: 0.2% BSA (Sigma, USA) in 1× PBS, filtered.
6. Ethanol solution: 70% Solution in water, filtered.
7. Materials for microfluidic channels: polydimethylsiloxane (PDMS) kit (Dow Corning, USA), glass slide.
8. Equipment for making enclosed microfluidic channel: plasma oxidizer (SAMCO, Switzerland).
9. Equipment for fluid perfusion: syringe pumps (Harvard Apparatus, USA).
10. Tubings and fittings: hole puncher (Technical Innovations, USA), Tygon® tubings (OD 1/32″, Saint-Gobain, France), gas permeable tubing (OD 1/8″, Dow Corning, USA), steel tubings (New England Small Tube Company, USA), two-way valve with a luer connection (Cole-Palmer, USA), valves, four-way valves, ferrules, and nuts (Upchurch Scientific, USA).

3. Methods

3.1. Fabrication of Microfluidic Channels by Replica Molding Using PDMS

All procedures are carried out at room temperature unless otherwise specified.

1. Mix thoroughly the PDMS base and the curing agent at the ratio of 10:1 (weight) according to the manufacturer.
2. Degas the PDMS–curing agent mixture by placing it into a vacuum chamber for approximately 30 min (see Note 5).
3. Place the fabricated silicon wafer into a petri dish with an appropriate size (see Note 6).
4. Gently pour the PDMS–curing agent mixture onto the wafer until it is completely covered and the preferred thickness of the mixture in the petri dish is 3–5 mm (see Note 7).
5. Place the petri dish containing both silicon wafer and PDMS–curing agent mixture into an oven and bake under 70–80°C for two hours (see Note 8).
6. When PDMS is cured, gently remove it from the wafer and the pattern of the microfluidic channel is obtained on the PDMS slab (see Note 9).

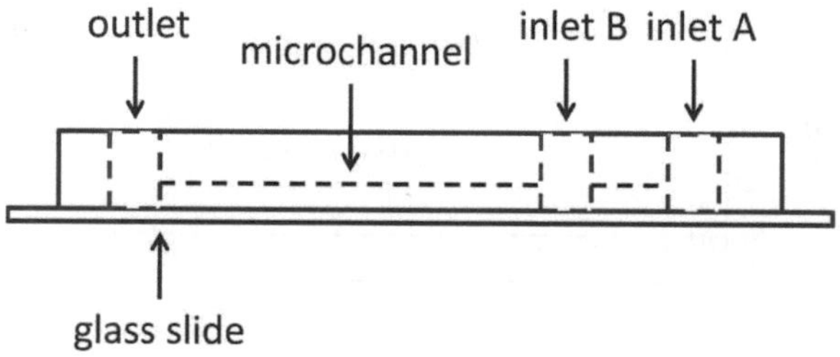

Fig. 2. Side view of the microfluidic channel bonded to a glass slide. Holes at the inlets and the outlet are punched to facilitate fluidic connections.

3.2. Surface Treatment of PDMS Slab for Irreversible Bonding to Glass Slides

In order to culture cells in microfluidic channels, the PDMS slabs have to be permanently bonded to clean glass slides to effect enclosed devices. All procedures are carried out at room temperature unless otherwise specified.

1. Punch holes at the two inlets and the outlet of the microfluidic channel in PDMS slabs to facilitate fluidic connection. It is recommended to use a sharp puncher (see Note 10).
2. The punched PDMS slabs are thoroughly rinsed with acetone, isopropanol, water and air-dried. Remove any dust that remains on the PDMS.
3. Soak the glass slides in 5 M of NaOH for 1–2 h and rinse them with running water. Blow dry and keep them in a clean, covered petri dish to avoid dust.
4. Put both the PDMS slab and the glass slide into the plasma oxidizer for 1 min at 125 W, 13.5 MHz, 50 sccm, and 40 mTorr (see Note 11).
5. Immediately lay the PDMS slabs on top of the glass slides and gently press to ensure irreversible chemical bonding. The microfluidic channels are ready to use (Fig. 2).
6. Sterilize all bonded microfluidic channels by autoclaving before cell culture.

3.3. Setting Up the Microfluidic Connections

In order to be able to culture mammalian cells in the microfluidic channels, fittings and connectors are to be used for fluidic handlings:

1. Cut the Tygon® tubings into 10-cm pieces.
2. Connect the tubings to ferrules and nuts so that they can be connected to valves for fluidic handling (see Note 12).
3. Cut 50 cm of silicon gas permeable tubing (Fig. 3a).
4. Connect the microfluidic channels to the tubings via small steel tubings (Fig. 3b).
5. Cut off a 22-G needle until it is 4–5 mm in length (Fig. 3c).

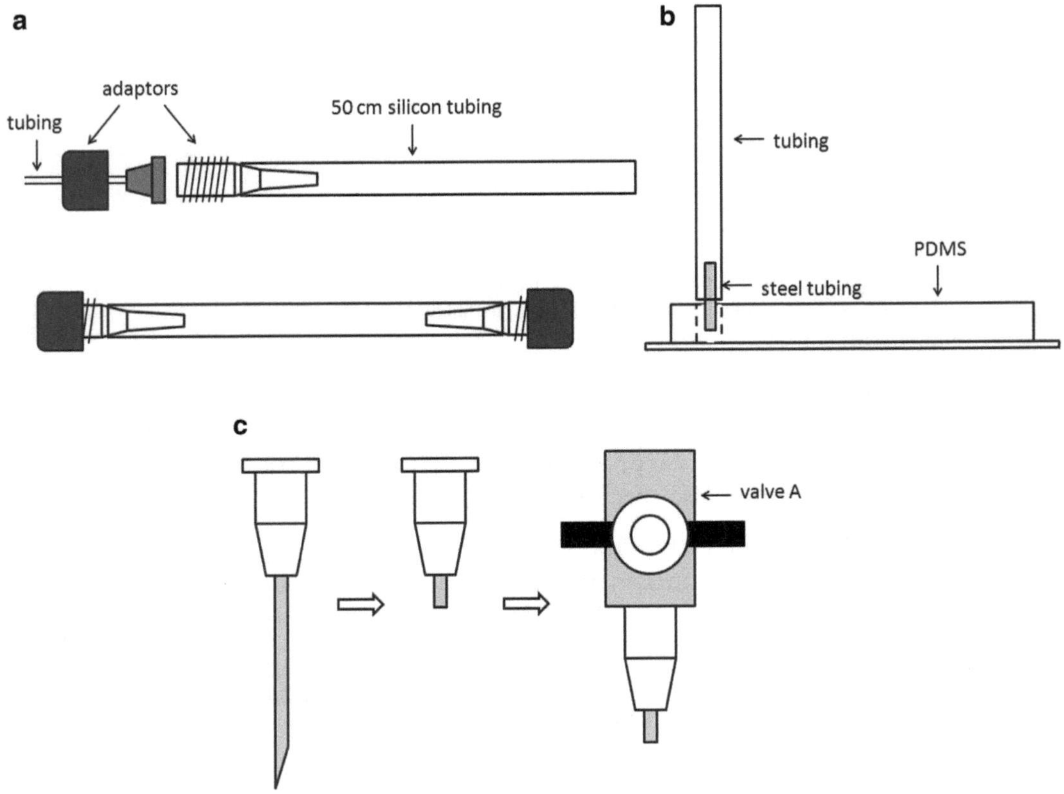

Fig. 3. Schematic illustration of tubing connections and the cell reservoir. (**a**) The normal Tygon® tubings are connected to the gas permeable tubings via adaptors and ferrules. Both ends of the gas permeable tubings are connected to adaptors for fluid flow. (**b**) The small steel tubings are 1 cm in length and work as needles. They can be directly plugged into the holes in the PDMS without causing leakage. (**c**)The normal 22 G needle is cut off until it is 4–5 mm in length. The needle is then coupled to a two-way valve (valve A) and they work as a cell reservoir.

6. Couple the needle to the two-way valve with a luer connection as a cell reservoir (Fig. 3c).
7. Sterilize the whole setup and all the components by autoclaving.

3.4. Cell Maintenance

1. Cell lines are purchased from ATCC, USA. In this case, HepG2 and MCF-7 are used as examples.
2. Cells are maintained in 12–14 mL of culture medium at 37°C with 5% CO_2. A cell culture flask with the surface area of 75 cm^2 is used.
3. Culture medium is changed 2–3 times a week to maintain cell growth. Warm up fresh medium to 37°C in a water bath and remove old culture medium from the flask using a sterile pipette (see Note 13).
4. Rinse the flask with sterile 1× PBS.
5. Add 12–14 mL of fresh medium and keep them at 37°C with 5% CO_2 (see Note 14).
6. Cells have to be subcultured when they reach 80% confluency (see Note 15).

7. Warm up culture medium and 0.25% Trypsin–EDTA solution to 37°C.
8. Remove old culture medium from the flask using a sterile pipette and rinse the flask with sterile 1× PBS.
9. Add 3 mL of 0.25% Trypsin–EDTA solution to the 75-cm^2 flask. Incubate at 37°C for several minutes until all cells detach from the bottom.
10. Add 4–5 mL of fresh culture medium into the flask and transfer all cell suspension to a centrifuge tube.
11. Spin down the cells at 300 ×*g*, 4°C for 3 min. Discard the supernatant and replenish with fresh culture medium.
12. Resuspend the cell pellet gently with a micropipette. Seed the cells into a 75-cm^2 flask at a density of ~6×10^5 cells/25 cm^2. Add more fresh culture medium until it is 12–14 mL.

3.5. Assembly and Priming of the Microfluidic Channels

Priming the microfluidic channel prior to cell seeding is important for cell culture. Microfluidic devices are prone to air bubbles which impede cell viability and functionalities. The purpose of priming is to completely remove air in the channel as well as the fluidic connections. By priming with 0.2% BSA, the protein can be coated onto the surface of the PDMS channels which helps perfusion cell culture.

1. Degas 70% ethanol and 0.2% BSA in a vacuum chamber for 15 min.
2. Connect inlet A of the microfluidic channels to the tubings as shown in Fig. 4. Tighten the nuts to avoid leakage.

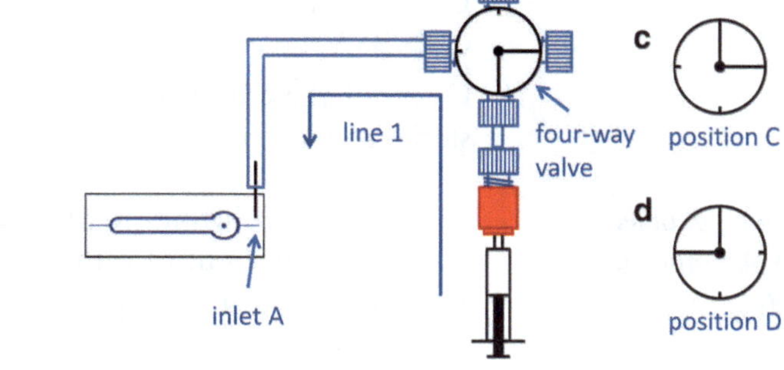

Fig. 4. Schematic illustration of connecting *line 1* to inlet A. (**a**) *Line 1* is used for priming the microfluidic channel. *Line 2* is used later for perfusion culture. (**b**) The four-way valve has to be kept in position B to get *line 1* through. (**c**, **d**) The four-way valve has two more positions to control different lines.

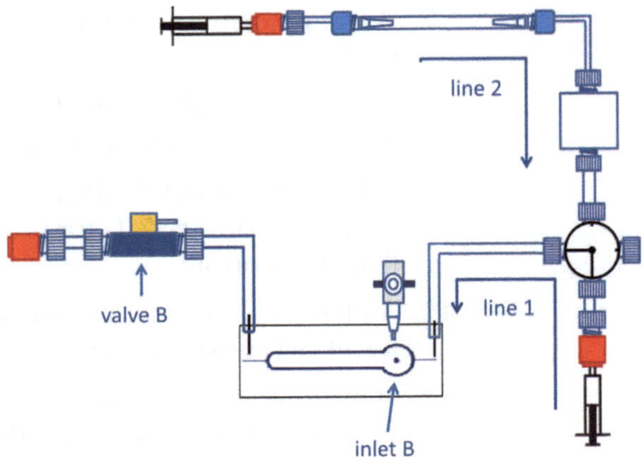

Fig. 5. Connecting valve B to the outlet and the cell reservoir to inlet B. While priming the entire microfluidic channel from *line 1*, the four-way valve is kept in the position as shown here.

3. Gently prime line 1 with 70% ethanol to remove air from the microfluidic channel (see Note 16). Liquid will come out from inlet B and the outlet.
4. Connect the outlet to the tubings and prime the microfluidic channel further with 70% ethanol. More liquid will come out from inlet B and the tubing from the outlet (Fig. 5).
5. Gently prime the cell reservoir (as shown in Fig. 3c) with 70% ethanol and remove all the air completely.
6. Gently rinse the cell reservoir with 0.2% BSA to replace ethanol thoroughly.
7. Connect the cell reservoir to the microfluidic channel via inlet B as shown in Fig. 5.
8. Gently prime the microfluidic channel with 0.2% BSA from line 1 via inlet A to the outlet to remove remaining ethanol. Repeat this step three times.
9. Gently prime line 2 with 70% ethanol followed by rinsing with 0.2% BSA three times.
10. Close all the valves and leave the entire fluidic setup under sterile conditions for one hour.

3.6. Seeding of Cells into the Microfluidic Channels

1. Connect a 1-mL syringe to the outlet via the luer connection and place the syringe onto a withdrawal syringe pump (Fig. 6a).
2. Move the microfluidic channel under a microscope for observation (Fig. 6a).
3. Suspend 9–10 × 10^6 cells in 1 mL of 3.0 mg/mL collagen.
4. Transfer 100 μL of the cell suspension into the cell reservoir and open valve A.

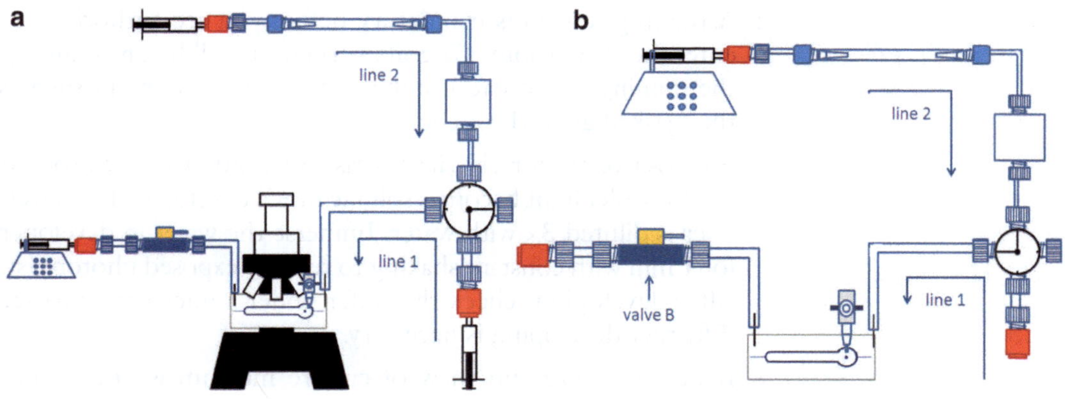

Fig. 6. Schematic representation for cell seeding and perfusion culture. (**a**) The outlet is connected to a syringe and placed onto a withdrawal pump for cell seeding. During cell seeding the four-way valve is kept in the position as shown here. Cells are withdrawn into the microfluidic channel and the cell seeding process is observed through a ×10 objective. (**b**) The syringe and the pump are removed from the outlet and valve B is switched open immediately after cell seeding. Fresh medium is infused from the syringe via *line 2* while the four-way valve is kept in the position as shown here. Valve B is kept open so that medium can come out from the outlet. *Line 1* is not in use during perfusion cell culture.

5. Open valve B and switch on the pump. Adjust the withdrawal rate to ~0.02 mL/h so that the cell suspension slowly enters the microfluidic channel from the cell reservoir.
6. Observe via a 10× objective and gently tap the microfluidic channel if necessary to prevent cells from clogging.
7. When the microfluidic channel is fully packed with cells, close both valve A and B. Disconnect the syringe from the outlet and open valve B.
8. Slowly infuse fresh culture medium from line 1 at a flow rate of 0.03 mL/h to remove excessive cells from the two flanking channels (Fig. 6b).

3.7. Perfusion Culture of Cells in Microfluidic Channels

1. Transfer the entire experimental setup onto a 37°C heating plate or alternatively into a 37°C incubator (see Note 17).
2. Infuse cell culture medium from line 2 at a flow rate of 0.03–0.05 mL/h (see Note 18) (Fig. 6b).
3. Collect the perfusate from the outlet if further analysis is needed.

4. Notes

1. The photoresist is a viscous liquid. It is spin-coated onto the wafer to produce a uniform layer. Dispense an appropriate amount of photoresist on the wafer and then spin at 1300 × *g* for 30 s. The coated wafer is then baked at 90°C for 1 min.

2. A training session is mandatory before you are authorized to enter the clean room. Careful instructions will be given during the training on how to use the clean-room equipment such as the mask aligner, the etcher, etc.
3. For positive photoresist (in this case AZ4620), the area exposed to ultraviolet light becomes soluble in AZ developer. The developer is diluted 3× with water. Immerse the wafer in developer for 1 min with constant shaking to remove exposed photoresist. After developing, check the wafer under a microscope to see if further developing is necessary.
4. Repeated freeze and thaw of culture medium is not recommended. It is suggested to aliquot culture medium into smaller containers. Store them at 4°C, protected from light. They can be stored for 1 month.
5. In order to remove air bubbles from PDMS that may block the microscale structures on the microfluidic channel, well-mixed PDMS–curing agent should be thoroughly degassed before they are cured.
6. The petri dish works as a container for the silicon wafer. For a 4″ wafer, the recommended diameter for a petri dish is 15–20 cm.
7. A too thin layer of PDMS may cause difficulties to fluidic connection. In order to avoid excessive consumption of reagent, the PDMS should not be too thick either. Gently pour the PDMS–curing agent mixture on top of the wafer to avoid air bubble formation.
8. Make sure that the oven is carefully leveled before baking. It is recommended to use a glass petri dish instead of a plastic one under high temperature.
9. Use a tweezer to help remove PDMS from the wafer. Do not scratch the surface of the wafer.
10. A sharp puncher can greatly help the formation of holes with smooth edges which can efficiently decrease the chances of air bubble formation in the fluidic connections during long-term perfusion cell culture.
11. The purity of oxygen has to be high to obtain effective plasma oxidation. Therefore, after placing PDMS and glass slides into the plasma oxidizer, it is recommended to evacuate the chamber before switching on oxygen infusion.
12. Connect the tubings to ferrules and nuts as shown in the illustration to ensure tight fitting (Fig. 7).
13. For suspended cell culture, cells will be removed from the flask together with old culture medium. It is necessary to spin down the suspension at 1,000 rpm for 3 min and discard the supernatant. Resuspend the cell pellet in fresh culture medium.

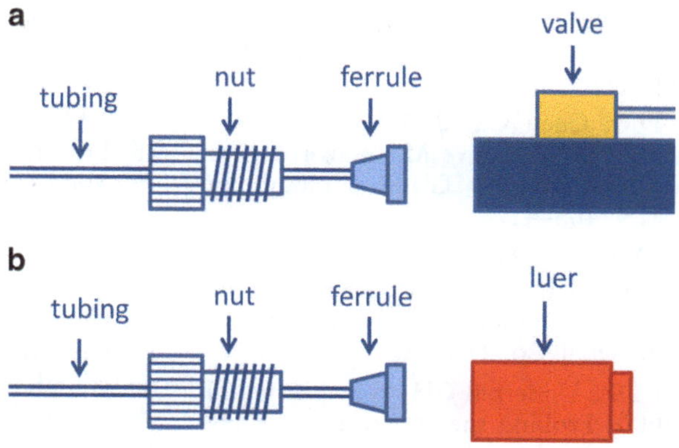

Fig. 7. Schematic representation for connecting the fittings. The nuts and ferrules are connected to the tubings in the way as shown here. The nuts can then be connected to (**a**) a valve or (**b**) a luer and the ferrules are used for tight fitting to prevent leakage.

14. Place the resuspended cells (see Note 13) back into the flask.
15. The subculture procedure is only applicable to adherent cell cultures.
16. The four-way valve has four different positions. When priming line 1, turn the four-way valve to the position as shown in Fig. 4a to remove air, followed by turning it to position B (Fig. 4b) to infuse liquid into the microfluidic channel.
17. If the entire perfusion culture is maintained on a 37°C heating plate without CO_2 supply, it is recommended to supplement the culture medium with 60 mM of HEPES buffer for pH equilibrium.
18. The gas-permeable silicon tubing allows efficient oxygen exchange. Therefore, culture medium should go from line 2 into the microfluidic channel. The bubble trap in line 2 helps remove large air bubbles in fresh culture medium. Line 1 is not in use during perfusion cell culture.

References

1. Karkkainen MJ, Makinen T, Alitalo K (2002) Lymphatic endothelium: a new frontier of metastasis research. Nat Cell Biol 4: E2–E5
2. Toma JG, Akhavan M, Fernandes KJL et al (2001) Isolation of multipotent adult stem cells from the dermis of mammalian skin. Nat Cell Biol 3: 778–784
3. Campenot RB, Eng H (2000) Protein synthesis in axons and its possible functions. Journal of Neurocytology 29: 793–798
4. Shofuda K-I, Hasenstab D, Shofuda T et al (2001) Control of Smooth Muscle Cell Function by Membrane-Type Matrix Metalloproteinases. Annals of the New York Academy of Sciences 947: 337–340
5. Bissell MJ, Barcellos-Hoff MH (1987) The influence of extracellular matrix on gene expression: is structure the message? Journal of cell science Supplement 8: 327–343
6. Metallo CM, Mohr JC, Detzel CJ et al (2007) Engineering the Stem Cell Microenvironment. Biotechnology Progress 23: 18–23
7. Lin C, Bissell M (1993) Multi-faceted regulation of cell differentiation by extracellular matrix. The FASEB Journal 7: 737–743
8. Badylak SF, Freytes DO, Gilbert TW (2009) Extracellular matrix as a biological scaffold material: Structure and function. Acta Biomaterialia 5: 1–13

9. Walker MR, Patel KK, Stappenbeck TS (2009) The stem cell niche. The Journal of Pathology 217: 169–180
10. Pirone DM, Chen CS (2004) Strategies for Engineering the Adhesive Microenvironment. Journal of Mammary Gland Biology and Neoplasia 9: 405–417
11. Torisawa Y-s, Shiku H, Yasukawa T et al (2005) Multi-channel 3-D cell culture device integrated on a silicon chip for anticancer drug sensitivity test. Biomaterials 26: 2165–2172
12. Albrecht DR, Underhill GH, Wassermann TB et al (2006) Probing the role of multicellular organization in three-dimensional microenvironments. Nat Meth 3: 369–375
13. Khademhosseini A, Langer R, Borenstein J et al (2006) Microscale technologies for tissue engineering and biology. Proceedings of the National Academy of Sciences of the United States of America 103: 2480–2487
14. Takayama S, Ostuni E, LeDuc P et al (2003) Selective Chemical Treatment of Cellular Microdomains Using Multiple Laminar Streams. Chemistry & Biology 10: 123–130
15. Lu H, Koo LY, Wang WM et al (2004) Microfluidic Shear Devices for Quantitative Analysis of Cell Adhesion. Analytical Chemistry 76: 5257–5264
16. El-Ali J, Sorger PK, Jensen KF (2006) Cells on chips. Nature 442: 403–411
17. Griffith LG, Swartz MA (2006) Capturing complex 3D tissue physiology in vitro. Nat Rev Mol Cell Biol 7: 211–224
18. Chung BG, Flanagan LA, Rhee SW et al (2005) Human neural stem cell growth and differentiation in a gradient-generating microfluidic device. Lab on a Chip 5: 401–406
19. Dittrich PS, Manz A (2006) Lab-on-a-chip: microfluidics in drug discovery. Nat Rev Drug Discov 5: 210–218

Chapter 9

Padlock Probes and Rolling Circle Amplification for Detection of Repeats and Single-Copy Genes in the Single-Cell Comet Assay

Sara Henriksson and Mats Nilsson

Abstract

Padlock probes and rolling circle amplification are techniques which can be used for detection of DNA sequences in situ with high specificity and high signal to noise. The single-cell gel electrophoresis assay is used to measure DNA damage and repair in cells. Here, we describe how padlock probes and rolling circle amplification can be used to detect DNA sequences within comet preparations.

Key words: Padlock probes, Rolling circle amplification, Comet assay, Single cell gel electrophoresis, DNA repair, DNA damage

1. Introduction

Padlock probes are single-stranded linear oligonucleotides that are ~90 nucleotides in length. Padlock probes are used to recognize specific DNA sequences by using ligation for detection. The two ends of the padlock probe are hybridized juxtaposed to a target DNA sequence, and upon perfect hybridization at the junction, the two ends are ligated forming a circular DNA molecule (1). This circularization reaction is highly specific, since it requires two hybridization events and a DNA ligation event. The DNA ligation reaction is specific enough to distinguish single nucleotide variations in the target sequence, which has been shown both in situ (2) and in solution, enabling massively parallel genotyping (3). Circularized probes can furthermore be specifically amplified using rolling circle amplification (RCA) (4, 5), generating long single-stranded concatemer copies of the probe sequence, which spontaneously coil into a

Sara Lindström and Helene Andersson-Svahn (eds.), *Single-Cell Analysis: Methods and Protocols*, Methods in Molecular Biology, vol. 853, DOI 10.1007/978-1-61779-567-1_9,

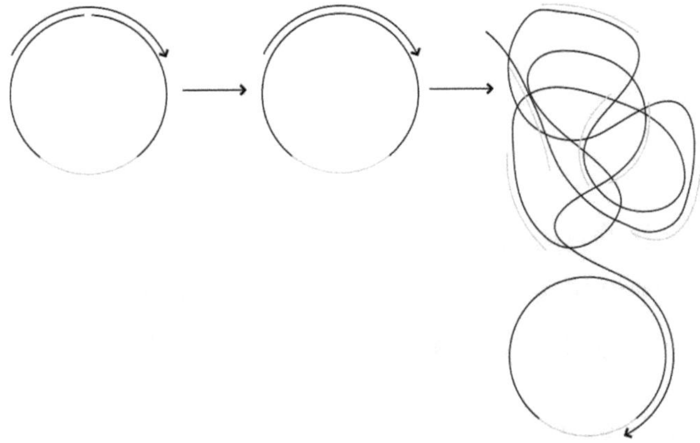

Fig. 1. Schematic illustration of the technique. The linear padlock probe is hybridized to a target molecule with the two ends juxtaposed. Upon perfect hybridization at the junction, the two ends are enzymatically ligated, forming a circular DNA molecule. The circularized padlock probe is then amplified with rolling circle amplification using the target molecule as a primer. The rolling circle product can be visualized by hybridization of fluorescently labeled oligonucleotides.

micron-sized blob of DNA that in a microscope appears as a bright fluorescent spot when labeled with fluorescence-labeled oligonucleotide probes (6, 7) (Fig. 1). Ligated padlock probes can also be amplified in situ via RCA using the target strand as a primer (8). The large rolling circle product (RCP) will then be anchored to the target molecule, thus preserving the localization of the detection signal. After labeling with fluorescent molecules, the rolling circle product is visible in a fluorescent microscope using 10× magnification or more. For quantitative analyses, the number of signals per cell can be counted using Olink image tool (9) or CellProfiler (10). In addition to detection of human genomic DNA sequences, the target-primed RCA technique has been applied for detection of viral DNA (11) and genotyping of mRNA (12).

The single cell gel electrophoresis assay, which is known as the comet assay, is an assay that is commonly used to measure DNA damage and repair. During the assay, cells are embedded in low melting point agarose on a microscope slide and treated with a toxin or radiation. After damage, the cells are lysed and then electrophoresed. Damaged DNA will then move through the gel, creating what looks like a comet tail, whereas intact DNA will remain in the area which used to contain the cell nuclei, which now looks like the head of a comet (13).

Comet assays are traditionally stained with a whole DNA stain such as DAPI or in some cases acridine orange but have also been detected with fluorescence in situ hybridization (FISH) (14). There are several advantages using padlock probes for detection of DNA in the comet assay compared to whole DNA staining and

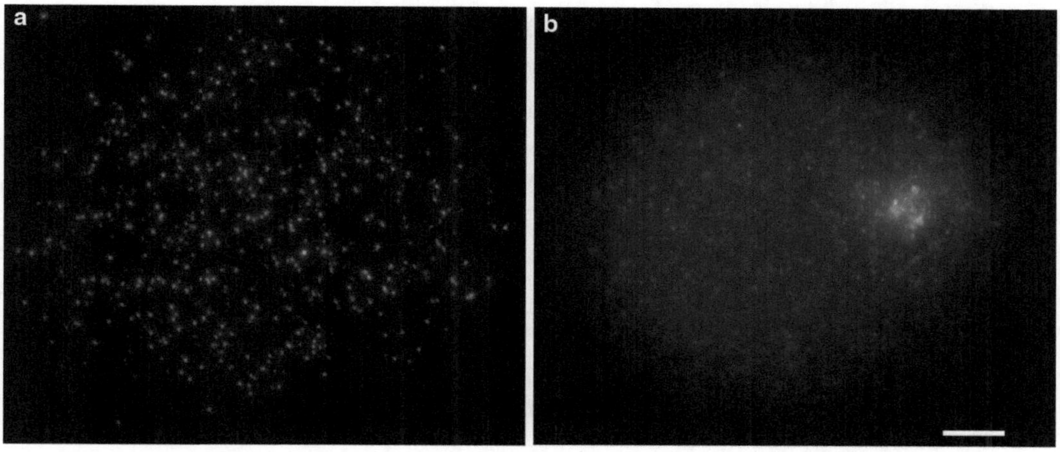

Fig. 2. Results from detection of *Alu* repeats in comet assays using padlock probes and rolling circle amplification. Rolling circle products are seen as dots in (**a**) and total DNA stain with DAPI is seen in (**b**). Scale bar is 10 μm.

FISH. Individual signals can be counted with padlock probes, thus eliminating background problems that arise with whole DNA stains. Padlock probes have short target sequences and can therefore detect smaller sequences than FISH. It is also a faster procedure and padlock probes and RCA can be used at 37°C and under low-stringent conditions which helps to preserve the agarose gel in the comet assay.

In this chapter, we describe how DNA sequences can be detected in comet preparations with padlock probes and rolling circle amplification. As an example, we have chosen to target the core sequence of Alu repeats, since these are present in high numbers in the genome (15) and can be used as a general marker for DNA damage in human comet preparations (16–18) (Fig. 2). For example single-copy gene locus sequences, we refer to (18).The procedure for performing comet assays is described elsewhere (19).

2. Materials

MQ H_2O should be used for all solutions and reaction mixes.

2.1. Pretreatment of Comet Preparation

1. Comet preparation on microscope slide.
2. 1× Phosphate buffered saline (PBS).
3. Weigh in 0.01 g of pepsin (SIGMA) and dilute it in 100 mL of 0.1 M HCl. Mix it thoroughly. Pour the solution into a Coplin jar and place a lid on top. Place the Coplin jar in a 37°C water bath. Leave the solution to heat for at least 20 min.
4. 70, 85, and 99.5% ethanol solutions: prepare and pour the solutions into Coplin jars with lids.

2.2. Phosphorylation of Padlock Probes

1. 10 U/μL T4 polynucleotide kinase (Fermentas).
2. 10× T4 polynucleotide kinase reaction buffer A (Fermentas).
3. 10 mM Adenosine triphosphate (ATP).
4. 10 μM padlock probe targeting Alu repeats: 5′-CTGGGATT ACAGG-CCTTTCCTACGACCTCAATGCACATGTTTGG CTCCTCTTC-GCCTCCCAAAGTG-3′. For example single-copy gene probe sequences, see ref. 18. The probe can be ordered with a 5′ phosphate or be phosphorylated according to Subheading 3.2.
5. Water bath set at 37°C.
6. Heating block set at 65°C.

2.3. Ligation

1. 1 U/μL T4 DNA ligase (Fermentas).
2. 10× T4 DNA ligase buffer (Fermentas).
3. 10 μg/μL Bovine serum albumin (BSA).
4. 2.5 M NaCl.
5. 2 μM Phosphorylated padlock probe.
6. 22 mm × 22 mm Coverslips.
7. Moist chamber, e.g., pipette tip box with paper wet with PBS.
8. Incubator set at 37°C.
9. Plastic beaker.
10. Tween-20 (Roche). Prepare a solution with 5% Tween-20 in H_2O, which will be used for buffers.
11. Stringent wash: 2× Saline sodium citrate (SSC) and 0.05% Tween-20. Pour 100 mL of the solution into a Coplin jar with lid and put it into a 37°C water bath. This solution can also be prepared in a volume of 1 L and the remaining solution can be kept in room temperature for future experiments.
12. Wash buffer: 0.15 M NaCl, 0.1 M Tris–HCl, and 0.05% Tween-20. Pour 100 mL into a Coplin jar. The buffer is also used after RCA and detection oligonucleotide hybridization. This buffer can be prepared in larger volumes, e.g., 1 L and the remaining buffer can be stored at room temperature and kept for future experiments.

2.4. Rolling Circle Amplification

1. Pap pen (ImmEdge hydrophobic barrier pen) (Vector laboratories).
2. 10 U/μL Phi29 DNA polymerase (Fermentas).
3. 10× Phi29 DNA polymerase buffer (Fermentas).
4. 10 mM dNTP.
5. 10 μg/μL BSA.
6. 50% Glycerol.

2.5. Hybridization of Detection Oligonucleotides

1. Prepare a 2× hybridization solution containing 40% formamide and 4× SSC. Store the solution dark in the freezer (see Note 1).
2. 10 μM detection oligonucleotide with a 5′ fluorophore and the same sequence as the middle part of the padlock probe: fluorophore-CCTCAATGCACATGTTTGGCTCC. The oligonucleotide should be ordered with HPLC purification.

2.6. Mounting and Microscopy

1. Prepare 100 ng/mL 4′,6-diamidino-2-phenylindole (DAPI) in Vectashield (Vector laboratories). Mix it carefully and store it dark at 4°C (see Note 2).
2. 24 mm × 40 mm Coverslips.
3. Nail polish.
4. Fluorescence microscope with at least 10× magnification and proper filters for the fluorophore of the detection oligonucleotide and for DAPI.

3. Methods

3.1. Pretreatment of Comet Preparations

1. Put the slides with the comets (see Note 3) in a Coplin jar with PBS. Remove the PBS and replace it with fresh PBS. Leave the slides for a few minutes.
2. Take the slides from the PBS and put them in the pepsin-HCl solution for 150″ (see Note 4). Then, immediately put them back in the Coplin jar with PBS. Replace the PBS with fresh PBS once.
3. Dry the slides by successively removing water by leaving them in 70, 85, and 99.5% ethanol for 3 min each. Then, leave the slides to air-dry standing up. Do not pour out the ethanol, since it can be used again later in the protocol.

3.2. Phosphorylation of Padlock Probe

1. Mix 1 μL T4 polynucleotide kinase with 5 μL 10× T4 polynucleotide kinase reaction buffer A, 5 μL ATP, 10 μL padlock probe and 29 μL H_2O in an Eppendorf tube. Incubate the tube in at 37°C for 30 min.
2. Inactivate the enzyme at 65°C for 5 min by putting the Eppendorf tube in a heating block. Keep the phosphorylated padlock probe in −20°C.

3.3. Ligation

1. Prepare a ligation mix by pipetting 2.5 μL T4 DNA ligase, 2.5 μL T4 DNA ligase buffer, 1.25 μL padlock probe, 2.5 μL ATP, 0.5 μL BSA, 2.5 μL NaCl, and 13.25 μL H_2O into an Eppendorf tube (see Note 5). Mix the solution by pipetting gently up and down. Include a technical negative control in your experiment where ligase is replaced with H_2O.

2. Pipette 20 μL ligation mix onto a clean 22 mm × 22 mm coverslip. Quickly flip the coverslip over so that the drop hangs under the coverslip. Put the coverslip with the drop onto the microscope slide with the comet so that the coverslip covers the gel and make sure there are no bubbles. Put the slide in a moist chamber and put it into a 37°C incubator for 1 h.
3. Remove the slide from the incubator when 1 h has passed and gently remove the coverslip by tapping it against the edge of a plastic beaker. If it does not come off easily dip it into the beaker with wash buffer and try again.
4. Put the slide in the Coplin jar containing 2× SSC and 0.05% Tween-20 for 5 min.
5. Then, transfer the slide to the Coplin jar with wash buffer for a few minutes.
6. Dry the slides by successively removing water by leaving them in 70, 85, and 99.5% ethanol for 3 min each. Then, leave the slides to air-dry standing up (see Notes 6 and 7).

3.4. Rolling Circle Amplification

1. Draw a circle on the gel using a pap pen when the gel is completely dry. The circle should have an inner diameter of ~1 cm.
2. Prepare a rolling circle amplification mix by pipetting 4.5 μL phi29 DNA polymerase, 4.5 μL phi29 DNA polymerase buffer, 1.13 μL dNTP, 0.90 μL BSA, 4.5 μL glycerol, and 29.48 μL H_2O into an eppendorf tube. Mix the solution gently by pipetting up and down carefully avoiding bubbles.
3. Carefully pipette 40 μL mix into the circle of the pap pen without damaging the gel (see Note 8). Put the slide into a moist chamber in a 37°C incubator for 1 h (see Note 9).
4. Take the slide out from the incubator and put it in the Coplin jar with wash buffer for a few minutes.
5. Dry the slides by successively removing water by leaving them in 70, 85, and 99.5% ethanol for 3 min each. Then, leave the slides to air-dry standing up.

3.5. Hybridization of Detection Oligonucleotides

1. Prepare a detection oligonucleotide hybridization mix by pipetting 12.5 μL 2× hybridization solution, 0.63 μL detection oligonucleotide, and 11.88 μL H_2O into an Eppendorf tube.
2. Pipette 20 μL hybridization solution onto a clean coverslip. Quickly flip the coverslip over so that the drop hangs under the coverslip. Put the coverslip with the drop onto the microscope slide with the comet so that the drop is inside the circle of the pap pen and make sure there are no bubbles. Put the slide in a moist chamber and put it into a 37°C incubator for 20 min.
3. Remove the slide from the incubator and gently remove the coverslip by tapping it against the edge of a plastic beaker.

If it does not come off easily, dip it into the beaker with wash buffer and try again.

4. Put the slide in the Coplin jar with wash buffer for a few minutes.
5. Dry the slides by successively removing water by leaving them in 70, 85, and 99.5% ethanol for 3 min each. Then, leave the slides to air-dry standing up.

3.6. Mounting and Microscopy

1. Pipette 20 μL solution of DAPI in Vectashield into the circle of the pap pen. Cover the microscope slide with a 24 mm × 40 mm coverslip and seal the edges with nail polish (see Note 10).
2. Study the slides in a fluorescence microscope with at least 10× magnifications. Images can be analyzed with image analysis software such as Olink image tool or CellProfiler which can count the individual signals in each comet (see Note 11).

4. Notes

1. When stored dark in the freezer the 2× hybridization solution can be stored for several months.
2. We used alkaline comets for detection of DNA in the comet assay. There have been indications that the technique works on neutral comets as well (16).
3. DAPI in Vectashield can be prepared at larger volumes and kept cold and dark to be used at later experiments.
4. Pepsin is a critical step for the assay and the time should be optimized carefully and kept exactly the same for all slides in an experiment. Shorter pepsin treatment than optimal could result in no signals, whereas longer treatment might result in loss and dispersion of DNA.
5. Ligation can be performed with T4 DNA ligase at 37°C when detecting repeated sequences. For detection of singe copy targets, a thermostable ligase such as ampligase should be used to ensure specificity of the signals (18).
6. Drying of the slides with ethanol between the reactions is not necessary but helps to preserve the shape of the gel and to avoid the gel deattaching from the microscope slide.
7. After drying the slide with ethanol after a reaction the slide can be left dry overnight and the experiment can be resumed the following day.
8. RCA can be performed in a circle by an ImmEdge hydrophobic barrier pen or in the wells of a silicone gasket. The gasket will eliminate the risk of evaporation and leakage of the reaction

solution as well as defining the area of the reaction (17). RCA can also be performed under a coverslip, but this will decrease the sensitivity of the assay significantly.

9. The reaction time for phi29 DNA polymerase is not crucial. A shorter reaction time will result in smaller rolling circle product and a longer reaction time will result in a larger rolling circle product. Longer time thus generates stronger signals but limits the number of signals that can be resolved. However if the reaction dries out there will be a total loss of signals.
10. When the slides have been mounted they can be stored for several months before microscopy if they are kept dark and cold (4°C).
11. For counting rolling circle products with image analysis software, images should preferentially be taken as z-stacks to avoid rolling circle products not being counted caused by out of focus.

References

1. Nilsson, M., Malmgren, H., Samiotaki, M., Kwiatkowski, M., Chowdhary, B. P., and Landegren, U. (1994) Padlock probes: circularizing oligonucleotides for localized DNA detection, *Science 265*, 2085–2088.
2. Nilsson, M., Krejci, K., Koch, J., Kwiatkowski, M., Gustavsson, P., and Landegren, U. (1997) Padlock probes reveal single-nucleotide differences, parent of origin and in situ distribution of centromeric sequences in human chromosomes 13 and 21, *Nat Genet 16*, 252–255.
3. Hardenbol, P., Baner, J., Jain, M., Nilsson, M., Namsaraev, E. A., Karlin-Neumann, G. A., Fakhrai-Rad, H., Ronaghi, M., Willis, T. D., Landegren, U., and Davis, R. W. (2003) Multiplexed genotyping with sequence-tagged molecular inversion probes, *Nat Biotechnol 21*, 673–678.
4. Fire, A., and Xu, S. Q. (1995) Rolling replication of short DNA circles, *Proc Natl Acad Sci USA 92*, 4641–4645.
5. Baner, J., Nilsson, M., Mendel-Hartvig, M., and Landegren, U. (1998) Signal amplification of padlock probes by rolling circle replication, *Nucleic Acids Res 26*, 5073–5078.
6. Lizardi, P. M., Huang, X., Zhu, Z., Bray-Ward, P., Thomas, D. C., and Ward, D. C. (1998) Mutation detection and single-molecule counting using isothermal rolling-circle amplification, *Nat Genet 19*, 225–232.
7. Blab, G. A., Schmidt, T., and Nilsson, M. (2004) Homogeneous detection of single rolling circle replication products, *Anal Chem 76*, 495–498.
8. Larsson, C., Koch, J., Nygren, A., Janssen, G., Raap, A. K., Landegren, U., and Nilsson, M. (2004) In situ genotyping individual DNA molecules by target-primed rolling-circle amplification of padlock probes, *Nat Methods 1*, 227–232.
9. Allalou, A., and Wahlby, C. (2009) BlobFinder, a tool for fluorescence microscopy image cytometry, *Comput Methods Programs Biomed 94*, 58–65.
10. Lamprecht, M. R., Sabatini, D. M., and Carpenter, A. E. (2007) CellProfiler: free, versatile software for automated biological image analysis, *Biotechniques 42*, 71–75.
11. Henriksson, S., Blomstrom, A. L., Fuxler, L., Fossum, C., Berg, M., and Nilsson, M. (2011) Development of an in situ assay for simultaneous detection of the genomic and replicative form of PCV2 using padlock probes and rolling circle amplification, *Virol J 8*, 37.
12. Larsson, C., Grundberg, I., Soderberg, O., and Nilsson, M. (2010) In situ detection and genotyping of individual mRNA molecules, *Nat Methods 7*, 395–397.
13. Ostling, O., and Johanson, K. J. (1984) Microelectrophoretic study of radiation-induced DNA damages in individual mammalian cells, *Biochem Biophys Res Commun 123*, 291–298.
14. Santos, S. J., Singh, N. P., and Natarajan, A. T. (1997) Fluorescence in situ hybridization with comets, *Exp Cell Res 232*, 407–411.
15. Rudiger, N. S., Gregersen, N., and Kielland-Brandt, M. C. (1995) One short well conserved

region of Alu-sequences is involved in human gene rearrangements and has homology with prokaryotic chi, *Nucleic Acids Res 23*, 256–260.

16. Shaposhnikov, S., Larsson, C., Henriksson, S., Collins, A., and Nilsson, M. (2006) Detection of Alu sequences and mtDNA in comets using padlock probes, *Mutagenesis 21*, 243–247.
17. Shaposhnikov, S., Azqueta, A., Henriksson, S., Meier, S., Gaivao, I., Huskisson, N. H., Smart, A., Brunborg, G., Nilsson, M., and Collins, A. R. (2010) Twelve-gel slide format optimised for comet assay and fluorescent in situ hybridisation, *Toxicol Lett 195*, 31–34.
18. Henriksson, S., Shaposhnikov, S., Nilsson, M. & Collins, A. (2011) Study of gene-specific DNA repair in the comet assay with padlock probes and rolling circle amplification, *Toxicol. Lett 202*, 142–147.
19. Dhawan, A., Anderson, D., (Ed.) (2009) *The Comet Assay in Toxicology*, RSC Publishing.

Chapter 10

Droplet Microfluidics for Single-Cell Analysis

Eric Brouzes

Abstract

This book chapter aims at providing an overview of all the aspects and procedures needed to develop a droplet-based workflow for single-cell analysis (see Fig. 10.1). The surfactant system used to stabilize droplets is a critical component of droplet microfluidics; its properties define the type of droplet-based assays and workflows that can be developed. The scope of this book chapter is limited to fluorinated surfactant systems that have proved to generate extremely stable droplets and allow to easily retrieve the encapsulated material. The formulation section discusses how the experimental parameters influence the choice of the surfactant system to use. The circuit design section presents recipes to design and integrate different droplet modules into a whole assay. The fabrication section describes the manufacturing of microfluidic chip including the surface treatment which is pivotal in droplet microfluidics. Finally, the last section reviews the experimental setup for fluorescence detection with an emphasis on cell injection and incubation.

Key words: Single-cell analysis, Droplet microfluidics, Soft lithography, PDMS, Microfluidics, Microfluidic design, Drop-based microfluidic, High throughput, Fluorescence detection, Screening, Droplet modules

1. Introduction

Droplet microfluidics compartmentalizes each assay in an aqueous microdroplet (1 pL–10 nL) surrounded by an immiscible oil. The technology presents a new paradigm for screening, providing increased throughputs, reduced sample volumes and single-cell analysis capabilities [1–5]. The droplet technology is enabled by the ability to generate and digitally manipulate droplets at very high-throughput (up to 5,000 per second), and by the development of surfactant systems that permits the production of very stable droplets that can be incubated off-chip and reintroduced intact in a microfluidic chip for analysis [6–9]. The droplet format provides physical and chemical isolations that eliminate cross-contamination,

Sara Lindström and Helene Andersson-Svahn (eds.), *Single-Cell Analysis: Methods and Protocols*, Methods in Molecular Biology, vol. 853, DOI 10.1007/978-1-61779-567-1_10, © Springer Science+Business Media, LLC 2012

as well as fast and efficient mixing of reagents. In addition, as a small number of cells can be analyzed in discrete droplets, this technology is particularly suitable for working with cells of limited availability such as stem cells or primary cells from patients.

Droplets can be manipulated and processed at very high throughput in a number of ways: they can be merged or fused [10–15], injected with reagents [16], incubated on-chip [17, 18], sorted [11, 19], and their content can be mixed [20, 21]. This led to numerous droplet-based manipulations of cells that include cell encapsulation [7, 9], true single-cell encapsulation [22, 23], cell freezing for cryopreservation [24], fluorescence interrogation for gene expression analysis [25, 26], on-chip cell staining [6, 27], cell electroporation [28–30], cell lysis [31–33], analysis of single-cell secretion by fluorescence [34], multiplex analysis of single-cell secretion [35], multiplex functional analysis of single-cells [36], enhanced detection of cell surface biomarkers [37], cell sorting [38, 39], and single-cell genetic analysis [32, 40, 41]. Importantly, droplet microfluidics is modular, and these modules can be integrated to design a whole workflow for cell screening [6] (see Fig. 10.1).

2. Materials

2.1. Formulation

Surfactants: Dimorpholinophosphate-PFPE (DMP-PFPE) to synthesize following protocol in ref. 7; Polyethyleneglycol-PFPE (PEG-PFPE) to synthesize following protocol in ref. 8 or from Raindance Technologies; Ammonium salt of carboxy-PFPE (Carboxy-PFPE) to synthesize following protocol in ref. 42. Oils: FC-40, FC-3283 and HFE-7500 from 3M (USA). Destabilizer from Raindance Technologies (Lexington, USA). Nonexhaustive list of additives (for the aqueous or oil phase) that can be tested: PEG8000, PEG600, methyl cellulose of different viscosities; BSA, Serum, Spermine, chitosan-PEG (amine and amide); Pluronic F-68, Pluronic L-64, and Tetronic 1307 from BASF (Germany); Zonyl-FSO and Zonyl-FSN from Dupont (USA). Live/dead stains (examples: Calcein-AM and Sytox Orange from Invitrogen, USA). Cell culture. Assay stains. Fluorescence microscope.

2.2. Circuit Design

Computer- Aided Design (CAD) software: Macromedia Freehand (Adobe, USA), AutoCAD and AutoCAD LT (AutoDesk, USA), Corel Draw (Corel, USA), Clewin (PhoeniX Software, Netherlands), Ledit (Tanner EDA, USA), LinkCad (LinkCad, USA), Layout Editor (http://www.layouteditor.net/, Germany), Unigraphics (Unigraphics Solutions, USA), Solidworks (Dassault Systemes, France), or DraftSight (Dassault Systemes, France) which is very similar to AutoCAD but free. This list has been compiled from a poll

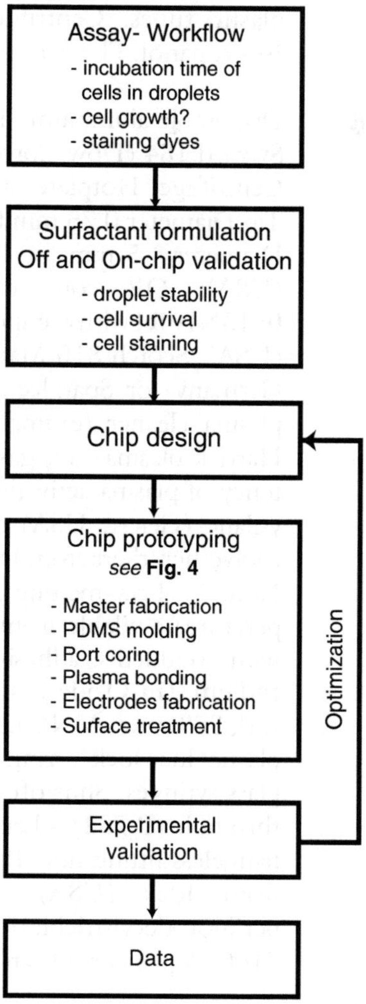

Fig. 1. Workflow for the development of a droplet-based assay for single-cell analysis.

conducted at the discussion group "Lab on a chip and Microfluidic Devices (aka Microfluidics)" on http://www.LinkedIn.com.

2.3. Master Fabrication

Printed mask mylar at 20,000 dpi (CADART, Bandon, OR) or chromium mask. SU8 photoresist and SU8 developer (MicroChem Corp., USA). Silicon wafer P(100) (ex: Test Grade Si 3″ wafer P(100) SSP (UW3P100) from University wafers). Spin-coater (examples: Laurell/WS-400B-6NPP-LITE or Headway Research/PWM32-PS-R790). Aligner/illuminator (examples: OAI- Hybralign Series 200 or Mask aligner Karl Suss/MJB-3). Profilometer (examples: Detak 150 Stylus Profilometer, Tesa Technology Visio 300, or noncontact profilometer Zygo New View 6000). (tridecafluoro-1,1,2,2,-tetrahydrooctyl)-1-trichlorosilane (TFOCS) (United Chemical Technology, USA). Smooth Cast 310 (Smooth-On Inc, USA). Hot plates. Noncontact thermometer to check temperature of hot plates. 1/32″ stainless steel dowell pins. 50-mL centrifuge plastic tubes. Centrifuge. Fridge. Desiccator jar and vacuum. Isopropanol. Flat borosilicate slide.

2.4. Microfluidic Chip Assembly

Disposable aluminum dish or aluminum foil. Nitrogen air. PDMS-Sylgard 184 (Dow Corning, USA). 50-mL centrifuge plastic tubes. Centrifuge. Hotplate or oven. Biopsy punchers: Harris Uni-Core, Tip Diameter 0.75 mm for 1/32″ tubing and Harris Uni-Core, Tip Diameter 1.5 mm for 1/16″ tubing available from Ted Pella Inc. (USA); OR—core drills CR0350275N20R4 and CR065-0515N16R4 with the appropriate pin vise from Technical Innovations (USA). Scotch 810 Magic Tape. Hellmanex III (Hellma Analytics, Germany) or Sparkleen (Fischer Scientific, USA) or acetone. O_2 plasma cleaner (examples of suppliers: Technics, Glow research, Harrick plasma). O_2 gas supply, vacuum sensor to improve consistency of plasma activation. Hot plate. 3-mercaptopropyltrimethoxysilane (Gelest, USA) diluted 1:10 in acetonitrile. 1/16″ PEEK (polyetheretherketone) tubing (0.04″ or 1 mm inner diameter) (see Note 1). Low-melting solder (Cerrolow-117, 47°C melting temperature) available from Mc Master Carr (USA). Copper foil tape with conductive adhesive (available from Mc Master Carr, USA). Indium Tin Oxide-coated glass slides (ITO) (Delta Technologies Ltd, Stillwater, MN). Electric wires. Wire stripper. 1-mL disposable plastic luer-lock syringes. Electrical multimeter. Luer-lock 5-mL glass syringes. Shut off valve for 1/16″ tubing: 1/4-28, 0.02 inch thru hole (P-732) + Female Luer ¼-28 Male tefzel (P-624) + super flangeless fitting nut (P-251X) + super flangeless ferrules (P-248X) from Idex (USA). 1/32″ PEEK tubing. 1H,1H,2H,2H-perfluorodecyl-trichlorosilane (Gelest, USA) 1% weight in HFE 7100. Aquapel applicator (PPG Industries, USA).

2.5. Experimental Setup

Pressure system (examples of suppliers: Fluigent, France; Dolomite, UK). Syringe pumps (examples of suppliers: Cetoni, Germany; Harvard Apparatus, USA; New Era, USA; Chemyx, USA; Cole Parmer, USA; Dolomite, UK). Inverted microscope with objectives. Photomultipliers, filters, dichroic mirrors, aperture, laser, camera. Labview graphical programming environment (National Instrument, USA). Stroboscope (example: X-strobe fromPerkin Elmer, USA). Fast shutter camera (example: XCD-V60 from Sony, Japan). Very high-speed cameras (examples of suppliers: Vision research, USA; or Photron, USA). EX-F1 high-speed camera (Casio, Japan). High-voltage module BXA 12579 (JKL Components, USA). DC power supply.

2.6. Cell Injection

1/32″ FEP (fluorinated ethylene propylene) tubing with a 400-μm inner diameter. 40-μm cell strainer (352340; BD Falcon). DNAse. Stirring bottle from Cetoni (Germany) or: Rare earth magnet for rotor (D44SH ¼″ diameter × ¼″ long; K&J Magnetic, USA); Nanoport (N126S; Idex, USA); Sample vial (Vial, Std Screw Thread, 12 × 32, 1.1 mL,Clear Glass, Tapered, Pulled Point. 1.1-STGV and Cap, Threaded, PP, 8-425, Open Top from SunSRI,

USA); 1/16″ OD × 1 mm ID PEEK tubing; Rare earth magnets for cell stirrer (D18- 1/16″ diameter × ½″ long; K&J Magnetic, USA); UV-curable glue (Loctite 3526 light cure adhesive); UV-curing lamp (examples: LED UV or Halogen UV curing guns from Edmund Optics, USA) and protective eyewear; Custom rubber gasket (8 mm OD × 4 mm ID).

2.7. Cell Collection and Incubation

2-µm disposable sample filter (B100; Idex). 1/32″ PEEK tubing. Female Luer ¼-28 Male tefzel (P-624; Idex) + PEEK Webbed union- PEEK (P-702) + super flangeless fitting nut (P-251X) + super flangeless ferrules (P-248X). 0.2-µm disk filter. Cell culture. 37°C, CO_2 incubator. PBS buffer. Petri dish.

2.8. General Experimental Tips

0.2-µm filter. 2-µm disposable sample filter (B100; Idex). Can of compressed air. Tweezers. 40-mm Cell strainer (352340; BD Falcon).

2.9. Microfluidic Interconnections

Razor blade or polymer tubing cutter (Idex, USA). Using fittings (see Note 1) from Idex (USA). To connect tubing to luer-lock syringes: for 1/16″ tubing-female Luer ¼-28 male tefzel (P-624) + PEEK Webbed union-PEEK (P-702). For 1/32″ tubing (see Note 2)-female Luer ¼-28 male tefzel (P-624) + PEEK Webbed union-PEEK (P-702) + super flangeless fitting nut (P-251X) + super flangeless ferrules (P-248X). Stopper to plug 1/32″ tubing: Plugs-red with white cap (P-771) (see Note 3). Shut off valve for 1/16″ tubing: 1/4-28, 0.02 inch thru hole (P-732).

Using fittings from VICI (USA): To connect 1/16″ tubing to 1/32″ tubing: 1/16″-1/32″ reducing union (ZRU1.5FPK). To connect 1/32″ tubing to 1/32″ tubing: 1/32″ internal union (ZU.5FPK). There are three different ways to adapt a union or fitting made for 1/16″ tubing to 1/32″ tubing: (1) Use super flangeless fitting nut (P-251X) + super flangeless ferrules (P-248X) when possible; (2) Use the 1/16″-1/32″ reducing union (ZRU1.5FPK) from Vici; (3) Use tubing sleeves (F-247x) from Idex.

3. Methods

3.1. Fluorescent Assays

Encapsulated cells are very difficult to wash (see Note 4). As a consequence, assays in droplets need to be based on fluorescent probes whose optical properties are altered during the assay (see Note 5), or based on a staining that develops inside the droplets. Changes in optical properties include the following: (1) Fluorescence enhancement [(a) DNA binding dyes that have very low fluorescence as free dyes and become bright upon DNA interaction (such as Sytox dyes, 7AAD, …) (6); (b) Probes that gain or lose fluorescence when cleaved, such as calcein-AM in which AM moieties quench the fluorescence of the green fluorescent calcein molecule (6); (c) Redox indicators such as alamarBlue (27)); (2)

Change in fluorescent polarization, (3) Change in local fluorescence concentration (34), and (4) Change of emission wavelength upon activation. Development of staining inside droplets include (5) fluorescence genetic reporters such as green fluorescent protein (GFP) for gene expression analysis (25, 26, 43) and (6) Enzymatic assay (37, 38, 44–46). Interestingly, cells can be labeled prior to encapsulation (47) and the fluorescent staining developed inside the droplet to take advantage of the concentrating of fluorescent molecules in a limited volume (37).

3.2. Experimental Parameters

1. Droplet stability is assured by surfactants that lower the surface tension between the aqueous and oil phases. In the case of single-cell analysis, the optimization of the surfactant/oil/additive system addresses three goals: (1) Droplet stability, (2) Cell survival, and (3) Optimal cell staining.
2. The following experimental parameters need to be considered for developing a droplet workflow for single-cell analysis: (1) Length of cell incubation in droplets, (2) Is cell growth inside droplet required? and (3) Fluorescent staining/reporter used.
3. Droplet stability depends on the surfactant/oil/aqueous buffer formulation; cell survival depends on the interactions between cells and the droplet interface; and cell staining inside droplet depends on the interaction between the dyes and surfactant molecules that may generate micelles that deplete the droplets of small molecules (48–53). Finally, cell growth depends on cell survival inside droplets and mainly on droplet size (6, 7).

3.3. Droplet Stability

Droplet stability is in part assured by the low surface tension between the aqueous and the oil phases due to the surfactant system. Different formulations of oil and surfactants have been developed, but systems based on fluorinated oil have the advantages of generating remarkably stable droplets even upon thermocycling (54). In addition, emulsions generated with these systems can be readily and easily broken with a chemical destabilizer (Raindance Technologies, USA) (7). Three perfluoropolyether-derived surfactants (PFPE surfactants) have been used in fluorinated droplet systems for analyzing cells: dimorpholinophosphate-PFPE (DMP-PFPE) (7), polyethyleneglycol-PFPE (PEG-PFPE) (8, 9), and ammonium salt of carboxy-PFPE (Carboxy-PFPE) (6, 42), and three different oils (FC-40, FC-3283, and HFE-7500) are routinely used (see Note 6). The following discussion sets up a guideline to optimize the formulation of a few systems and is not a definitive account of the different options available. This guideline is universal and useful to test the properties of different batches of surfactants and of new surfactants (55, 56). Importantly, different batches of surfactants can exhibit different properties and each batch should be tested.

Droplet stability is better tested by encapsulating the aqueous buffer on chip, collecting the droplet and reinjecting them (see Note 7). In general, HFE-7500 oil gives very stable droplets, but seems detrimental to cell survival when used with the Carboxy-PFPE surfactant (unpublished data). In that case, using a oil mixture FC-3283/HFE-7500: 90/10 weight seems to assure both stability and good cell survival (6). In conjunction with oil phases that contain enough HFE-7500, all surfactants generate pretty stable droplets.

3.4. Cell Survival

1. Long-term cell survival (longer than overnight) can be problematic for mammalian cells, but not for bacteria or yeast. Cell survival in droplets seems to depend solely on the interactions between the cells and the droplet interface whose chemical properties are dictated by the surfactant head. The PEG-PFPE surfactant proved to be biocompatible with mammalian cells in short-term incubations (6 h) (9). However, the PEG-PFPE surfactant does not preserve mammalian cell survival for long-term incubations (see Note 8), while the Carboxy-PFPE (with Pluronic F-68 as additive) and DMP-PFPE surfactants are biocompatible with mammalian cells in long-term incubations (6, 7).
2. Because cell survival depends on cell–interface interactions, it is possible to render a surfactant biocompatible by tailoring the interface properties using additives (6, 57). For example, U937 cells spread and lyse at the droplet interface when encapsulated with the Carboxy-PFPE surfactant alone but exhibit great survival rate when a cosurfactant (Pluronic F-68) is added to the aqueous phase (6). So, it is important to carry out a series of tests to find additives (see Note 9) able to tailor the interactions between the cells and the interface. This is done by seeding a cell solution on top of an oil layer in overlay experiments (see Fig. 2).

3.5. Overlay Experiments (6, 7)

1. In 96-well plate, dispense the fluorinated phase (about 50 μl), make sure that there is enough liquid to cover the whole bottom of the well.
2. Seed the cell solution (about 50,000 cells) on top of the fluorinated layer, adding your different additives. Incubate in a 37°C, CO_2 incubator.
3. Inspect the morphology of cells under a microscope during the course of several days. Cells should not interact with the interface and keep a rounded shape (see Note 10).

3.6. Quick Encapsulation Test (See Note 11)

1. The emulsion can be readily broken and cells be recovered and stained with live-dead fluorescent dyes (see Note 12) (7).
2. Encapsulate cells on a chip using formulations validated with the overlay experiment. Collect droplets in a microtube. Remove as much oil as possible by pipetting out the lower layer. Incubate in a 37°C, CO_2 incubator.

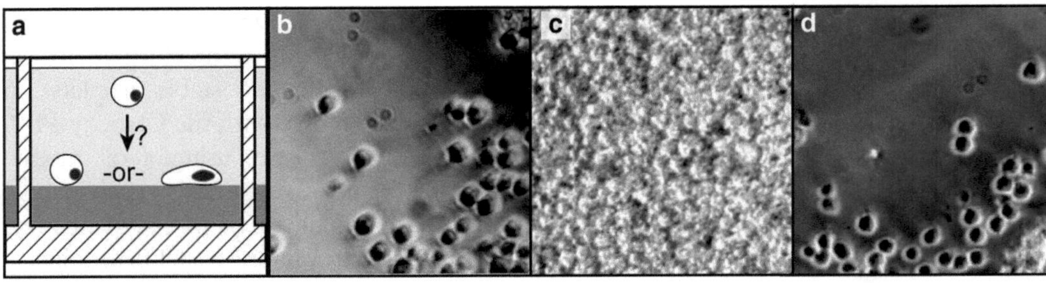

Fig. 2. Overlay assay to optimize system formulation. (**a**) Cells are seeded on top of an oil layer in a 96-well plate. Cell morphology is monitored over time to detect any interaction between the oil-buffer interface and the cells. (**b**) U937 cells seeded on top of FC-3283 oil only exhibit a rounded shape that indicates the absence of interactions. (**c**) A strong gel develops at the interface when U937 cells are seeded on top of FC-3283 oil containing 1% of carboxy-PFPE. (**d**) U937 cells seeded with 1% wt. Pluronic F-68 on top of FC-3283 oil containing 1% of carboxy-PFPE exhibit a rounded shape indicating that cell-interface interactions are minimal.

3. Break the emulsion with the destabilizer (Raindance Technologies): add slowly an equal volume of destabilizer to the emulsion. Let it incubate a few minutes, do not disturb (see Note 13); the emulsion should disappear and be replaced by an aqueous phase. Only when the emulsion is totally broken, add 200 μl of cell medium without disturbing the interface. Readily collect the upper layer and dispense in a well (see Note 14).
4. Stain with a live-dead assay to assess cell survival in your conditions.
5. Formulation needs to be optimized for each cell type and each batch of surfactant. Cell growth seems to depend solely on droplet size (7).

3.7. Assay Signal

One issue associated with surfactant is the depletion of small molecules from droplets through the formation of micelles (see Note 15). For example, the dye Sytox orange (Invitrogen) leaks out of droplets in a matter of seconds when encapsulated with the Carboxy-PFPE surfactant only. Additives can also tailor interactions between the small molecules and surfactant, permitting to slow down dye leakage as shown by using the cosurfactant Zonyl-FSO in the case of Sytox orange (6). Nevertheless, dye leakage is greatly reduced by the use of the PEG-PFPE surfactant which is a more satisfactory solution (6, 49). Often, the workflow can be designed so that the long-term cell incubation in droplets is conducted with one surfactant system, and the staining can be done on-chip by reinjecting the cell emulsion with the PEG-PFPE surfactant (6).

3.8. Fast Hand-Shake Emulsion Test for Assay Signal

1. A way to test for the compatibility of dyes with the surfactant is to conduct a fast hand-shake emulsion test (see Note 16).

2. Prepare a microtube containing 500 μl of oil and surfactant. Seed 50 μl cells on top without perturbing the interface; you want two clearly separated phases (see Note 17).
3. Add 50 μl of staining solution on top of the cells. Readily agitate the microtube by hand to mix the cell solution with the staining dyes and generate an emulsion.
4. Incubate and break the emulsion (as described earlier) with the destabilizer before transferring the cells into a well for inspection under a fluorescent microscope (see Note 18).

3.9. Circuit Design

1. Software and basic commands (see Note 19). Using AutoCAD, design your circuit by controlling the dimensions and keeping all the elements of the design linked with:
   - Line: controlling the location of the next point by using the @ command;
   - Offset command to duplicate object at a specific distance;
   - Trim, extend, fillet to link two lines together;
   - Snap mode is important to link a line or object to a design;
   - Copy as block to easily copy a block and explode to modify subcomponents of the block;
   - Regen all when zoom is blocked;
   - Measure distance command;
   - Make sure that the design is closed (see Note 20):
     - Convert the design to a polyline with the pedit command (pedit/multiple/join/fuzz distance: 1)
     - Hatch the design with the hatch command (solid pattern) making sure that you are working on the design layer
     - If the hatching fails the polyline is not closed; undo the hatching
     - In case the design is not closed, select a specific line of the design and check where the selection stops
   - Use different layers for depth 0, depth 1, ... and construction lines, wafer layout.
2. Encapsulation module. The nozzle of a flow-focusing device is essentially a cross made by two perpendicular channels. While the aqueous phase flows down the central channel, the two perpendicular channels facing each other are used to flow the oil phase to pinch the central streamline and trigger the formation of droplets (see Fig. 3a). Droplets form and flow through the fourth channel. No theoretical formula relates the nozzle dimensions to the droplet size (58). The droplet size depends on the nozzle size but also on the buffer composition and viscosity. As a starting point, we can use the following rule of thumb: when the width and depth of the nozzle are in the

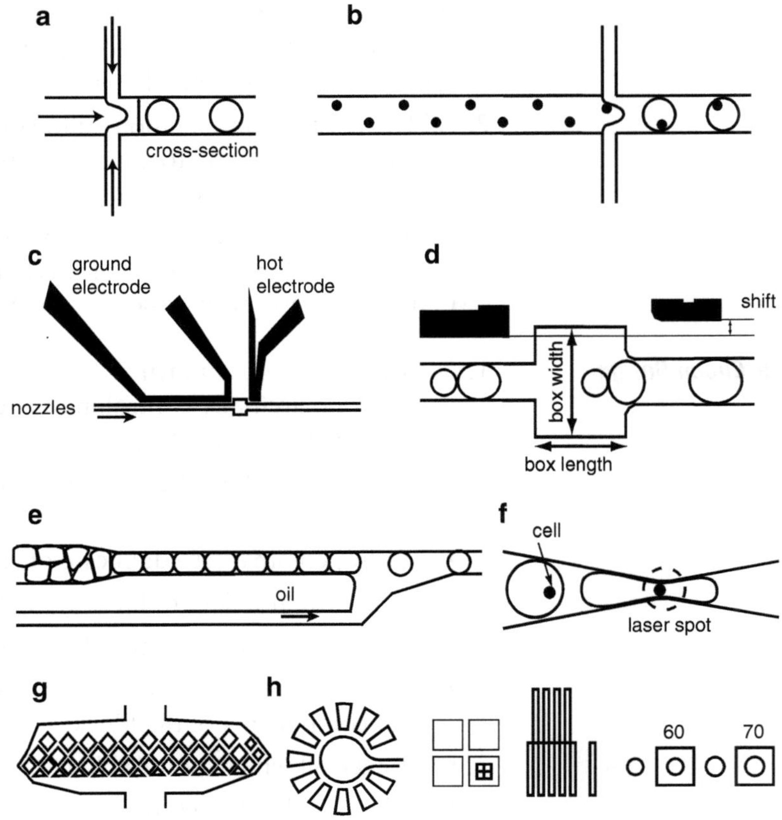

Fig. 3. Design of modules used to develop droplet-based workflow for single-cell analysis. (**a**) Nozzle for droplet generation. (**b**) Single-cell encapsulation, cells (represented as black dots) are first organized in a regular pattern in a long, high aspect ratio channel with dimensions close to the cell size. This channel is connected to a nozzle to encapsulate single-cells (adapted from ref. 23 reproduced by permission of The Royal Society of Chemistry). (**c**) Electrodes design for the merging module (box between the electrodes). (**d**) Details of the design of the merging module. (**e**) Reinjection module that gradually squeezes droplet in single-file to assure a constant spacing between reinjected droplets. (**f**) The detection module ensures a consistent excitation by aligning cells with the laser spot. (**g**) Example of a design for filtering contaminants. (**h**) Design structures that help design manufacturing and quality control, in order: a rosace-like structure for helping locate ports to be punched; an alignment mark for multilayer photolithography; resolution marks; and measurement marks to evaluate droplet size.

same dimension range, the droplet diameter is the arithmetic mean of the dimensions of the nozzle cross section: $\text{diameter}_{\text{drop}} = \sqrt{\text{width} \times \text{depth}}$ (see Note 21). Additionally, the nozzle dimensions control the volume fraction of oil over the aqueous phase. For instance, a low volume fraction can be obtained when the nozzle width is smaller than the nozzle depth (38).

3. Cell encapsulation following Poisson statistics. Because the process of loading cells into droplets is purely random, the distribution is dictated by Poisson statistics (7, 9). The Poisson

distribution depends only on the average number of events λ (here, the average number of cells per droplet). Single-cell works relying on Poisson encapsulation are usually conducted for λ lower than 0.3 (6, 7, 9, 27), which corresponds to a distribution where roughly one third of the droplets is empty, one third contains a single cell, and the remaining third contains more than one cell. This stochastic cell loading significantly reduces the number of usable droplets. To assure single-cell encapsulation, two recent methods can be used (22, 23). We describe the method that relies on inertial effects to pattern cells along a high aspect ratio channel (23).

4. True single-cell encapsulation by inertial effects. When cells are flowed into a high-aspect ratio channel with a cross section in the same range as the cell size, they order periodically in regular chains along the direction of the flow. To allow for true single-cell encapsulation, such a channel used to order cells is coupled to a nozzle (see Fig. 3b) (23). The ordering occurs in long channels and at fairly high-velocity so that the Reynolds particle number is closed to 1 (59). Because of the high velocity required in this design, it is difficult to integrate this module into a multistep device. In addition, it is practically difficult to encapsulate single cells in droplet much larger than 25–30 μm in diameter. To design such a system (60), carry out the following steps: (1) Calculate, based on the desired droplet size, the concentration of cells as if every droplet held exactly one cell; (2) Design a microfluidic device so that the cell length fraction defined as $l_c = \frac{6\phi \times \text{channel width} \times \text{channel height}}{\pi \times \text{cell size}^2}$, (where ϕ is the cell volume fraction) is close to 0.25. The microfluidic channel should be long (about 30 mm) and have a high aspect ratio with both dimensions smaller than ten times the cell size; (3) Flow aqueous phase containing cells (at a concentration of about 0.8–0.9 times the ideal concentration calculated in step 1) as fast as possible. Adjust oil flow rates to obtain droplets of desired size.

5. Merger module. The addition of material to droplets using surfactant relies mostly on electrocoalescence where a high-voltage high-frequency electric field is used to induce the merging of two droplets (see Note 22) (10). The microfluidic design uses the fact that droplets smaller than the channel size flow faster than droplets bigger than the channels. After droplets are inter-digitated through a Y-junction, they are left to catch up as pairs before entering a merging module (see Note 23). The spacing is dictated by the bigger droplets that create moving walls, and they are usually generated on-chip because droplet generation gives a more uniform spacing than reinjection. The role of the merging module is to close any gap between

droplets that form each pair, to slow down droplet pairs so that the effect of the electric field is more efficient (6, 61), and to avoid any electrowetting of droplets on the channel walls (see Fig. 3d). Design of the merging module (see Note 24): (1) The length and the width of the merging box are about twice the width of the incoming channel, assuming that the bigger droplets are about the size of the incoming channel (see Note 25). The outgoing channel is widened to accommodate the larger size of the merged droplet. (2) Electrodes are designed so that the "ground" electrode runs on a long distance along the incoming channel (see Fig. 3c). This is very important to avoid coalescence of the incoming emulsion. The "hot" electrode is narrower and placed after the merging box. The distance between the electrodes and the channel depends on the resolution that can be achieved during the fabrication of the master (see Subheading 3.12). It helps to move the "hot" electrode further away from the channel than the ground electrode to protect the incoming emulsion (about 20 μm) (38). The channels used to make the electrodes have both an inlet and outlet through which melting solder will be injected (see Subheading 3.12). They are designed in the same layer as the droplet channel for perfect positioning, but it helps to design the electrode sections closed to the ports in a deeper layer to facilitate solder injection.

6. Incubation line. The incubation times practically achievable on-chip range from 1 min to about 1 h. On-chip incubation is lengthened by using a long channel with a large cross section in order to increase the volume swept by the fluid (which is also key to reduce back-pressure, see Subheading 3.11) (see Note 26). The design can include some constrictions at regular intervals to continuously shuffle droplets in order to limit the dispersion of incubation times (18) (see Note 27).
7. Reinjection module. The reinjection module is designed so that droplets are reinjected with a spacing as constant as possible. It consists of a channel of gradually decreasing width to end up slightly narrower than a droplet so as to squeeze droplets into a single file (see Fig. 3e) (62). The oil solution used to respace the droplets is injected through a side channel which is inclined and wide to avoid splitting droplets upon reinjection (38) (see Note 28).
8. Detection module. To assure that encapsulated cells are consistently interrogated, the channel of the detection module is gradually constricted to align cells with the laser spot (see Fig. 3f) (6, 63). The detection module is fabricated in the shallowest layer because the depth of focus of the objective can also be limiting (z-constriction) (see Note 29) (6).

9. Filters. To manage the presence of contaminants (dust, fiber, ...) that can ruin experiments by clogging microfluidic channels, some online filters are added at the channel inlets whenever possible (oil, aqueous phases, reinjection lines). The design of a filter is based on a series of posts that create parallel and winding channels whose widths are the size of the smallest features of the whole design (see Fig. 3g) (see Note 30).

3.10. General Design Rules

The circuit design is constrained by the microfabrication process used to manufacture devices. Particularly, the smallest dimension and distance between two structures in the design is limited by the resolution allowed by the microfabrication process of the master (see Note 31). In addition, leave enough bonding surface between large structures such as incubation line (see Note 32). Finally, the chip design sets the layout for the tubing connections on your experimental station so try to optimize it during the conception. Add any design structure that would facilitate chip manufactuting, including the following: leaving enough space around the ports that will be punched; using a deep layer for the electrode sections closed to the ports; a rosace-like structure to help locate the ports to be punched (64); some alignment marks for multilayer lithography; some measurement marks to evaluate the droplet size by image analysis; and some resolution marks (see Fig. 3h, in order).

3.11. Modules Integration

Droplet microfluidics is a modular technology that allows for integrating different modules into a single device. The key for a successful integration is to (1) generate droplets of the right size, frequency and with the appropriate volume fraction and (2) to keep the back-pressure in the working range of the microfluidic devices and flow actuation system. Given back-pressure is in a reasonable range (below 15 psi), each module can be optimized separately. The design can be optimized by trial and error; PDMS stamping allows for the fabrication of several circuits on the same chip and a lot of design parameters can be tested in a short period of time (see Note 33).

The effect of back-pressure is twofold. First, high back-pressure increases the transition time to reach stable operation especially when it is due to a long incubation line. Second, the pressure drop may surpass the working range of the device which is estimated at three bars (300 kPa or 43.5 psi) for PDMS device bonded to a glass slide by plasma activation (65). Working in the higher range complicates manufacturing because of the difficulty to have consistent bonding by plasma activation, and results in a high number of devices failing by leakage or delamination. It is difficult to evaluate the fluidic resistance of a circuit for a 2-phase flow containing surfactant (66), but as a first approximation you can evaluate the pressure drop by considering the case of a 1-phase flow in a square channel for which the pressure drop is given by: $\Delta P = c\eta \frac{l}{wh^3} Q$, where Q is the flowrate; η the viscosity of the oil; l, w and h represent the length, width and height of the channel, and $c = 12\left[1 - \frac{192}{\pi^5}\frac{h}{w}\tanh\left(\frac{\pi w}{2h}\right)\right]^{-1}$ is a constant that depends on the channel geometry (w/h) (66). Typically, when computing the equation with the total flow rate you intend to use the pressure drop should be lower than 15 psi. In case the pressure drop is higher, you need to modify your design by increasing the thickness of some structures or shorten some narrow channels.

3.12. Chip Fabrication

1. The chips are PDMS-glass hybrid devices manufactured by soft-lithography (see Fig. 4) (67, 68).
2. Design your microfluidic circuit using a CAD software (see Subheading 2.2). Print or have mask printed at high resolution (20,000 dpi) on a mylar sheet (69) or a chromium mask if a better resolution is required (see Note 34).

3.13. Master Fabrication

1. The master is obtained by photolithography using the negative photoresist SU-8 (MicroChem, USA) on a silicon wafer (see Note 35). The following operations have to be conducted in a clean room (70).
2. Dehydrate the silicon wafer at 200°C on a hot plate for 10 min prior use to increase SU8 adhesion (see Note 36). Spin-coat the SU8 photoresist on the silicon wafer following the manufacturer's guidelines depending on the targeted thickness (see Notes 37 and 38). Soft-bake the wafer in a two-step incubation on hot plates at 65°C and 95°C to evaporate the photoresist solvent (see Note 39). Let the wafer cool down.
3. Sandwich the mask between the wafer and a flat borosilicate slide on the illuminator. Switch on the vacuum of the illuminator and bring the wafer in contact with the mask (see Note 40). Expose the photoresist (see Note 41).
4. After exposure, post-bake the wafer in a two-step incubation on hot plates at 65°C and 95°C to complete the polymerization process (see Note 42).
5. Development: the nonpolymerized photoresist is removed chemically using SU8 developer. Immerse the wafer in SU8 developer contained in a glass beaker for 6 min while constantly agitating (see Notes 43 and 44). Wash/rinse the master with isopropanol (see Note 45).
6. The resolution of the master can be readily assessed under a stereomicroscope by inspecting the critical structures of the design and the resolution marks.

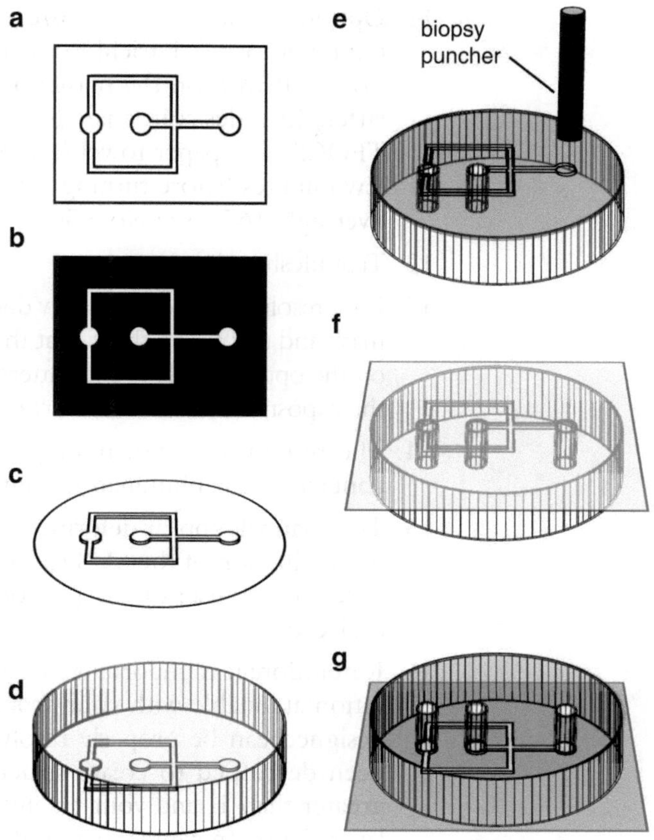

Fig. 4. Overview of chip manufacturing by soft-lithography. (**a**) Circuit design with CAD software. (**b**) Mask printing. (**c**) Master fabrication by photolithography. (**d**) PDMS molding. (**e**) Port coring. (**f**) Bonding of PDMS slab to glass slide by plasma activation. (**g**) The chip is ready for electrode fabrication and surface treatment.

7. Another SU8 layer can be added by following the same procedure. The only difference is the necessity of aligning the second mask using the alignment marks present on the first layer and on the mask (see Notes 46 and 47).
8. When done with the different layers hard bake the master on a hotplate at 120°C for 30 min.
9. Quality control: the layer thickness can be measured with a stylus profilometer, and the xy resolution with a noncontact comparator, or both with a 3D optical surface profilers.
10. Optional: transition step between two layers of different depths. Add a drop of photoresist at the location of the transition. Briefly spin the wafer or incubate it directly on a hot plate for the soft-bake. The photoresist will flow and form a meniscus between the two layers. Expose the master using a mask designed to define the transition step. Post-bake and then develop before doing the final hard bake (see Note 48).

11. Optional: Silanize the master with (tridecafluoro-1,1,2,2,-tetrahydrooctyl)-1-trichlorosilane (TFOCS) to limit the adhesion of PDMS on the master and increase the lifetime of the latter. In a desiccator jar, place the master and one drop of TFOCS on a paper towel (see Note 49). Apply vacuum for a few minutes before turning off and incubate between 2 h and overnight (67) (see Note 50).
12. Troubleshooting.
13. Poor resolution: most certainly due to a poor contact between the mask and the wafer; check that the emulsion of the mylar mask is on the opposite side of the pattern and is not facing the wafer; or the exposure is either too short or too long (see Note 51).
14. The resolution is not homogeneous across the master: poor contact; or the illumination source is not properly collimated.
15. The channels appear deformed or displaced: most likely due to poor adhesion of the SU8 to the wafer, need to dry/clean the wafer before use; or the photoresist is aged and needs to be replaced.
16. Resolution: it is important to have a good estimate of the resolution attainable with your process, so that the structures you designed can be properly resolved. The photoresist SU8 has been developed to create structures with aspect ratios much greater than 1, and your resolution should be better than the layer depth (meaning better than 50 μm for a 50 μm deep layer, and so on).

3.14. Plastic Master

1. Once the design has been validated and is expected to be used over a long series of experiments, it may be useful to make a more durable mold than SU8 on silicon wafer. Durable molds have been made in PDMS (71–73) or plastic materials (74, 75) and start from a PDMS mold made from the silicon master. These durable molds can be machined and integrate ports or reservoir.
2. The protocol for plastic mold is adapted from ref. 74 (see Fig. 5). The plastic master is manufactured using a commercially available two-part polyurethane plastic (Smooth Cast 310, Smooth-On Inc.). Store the two parts A and B at 4°C to increase the pot-life of the mixed polyurethane solution and to allow degassing.
3. Punch holes to create ports and clean the PDMS slab with a piece of tape (Scotch 810 Magic Tape) (see Subheading 3.16, step 1). Set the PDMS original in a PDMS container by seeding a thin layer of PDMS at the bottom of the container (see Fig. 5a). Add the pins you want transferred to the plastic mold (see Fig. 5b).

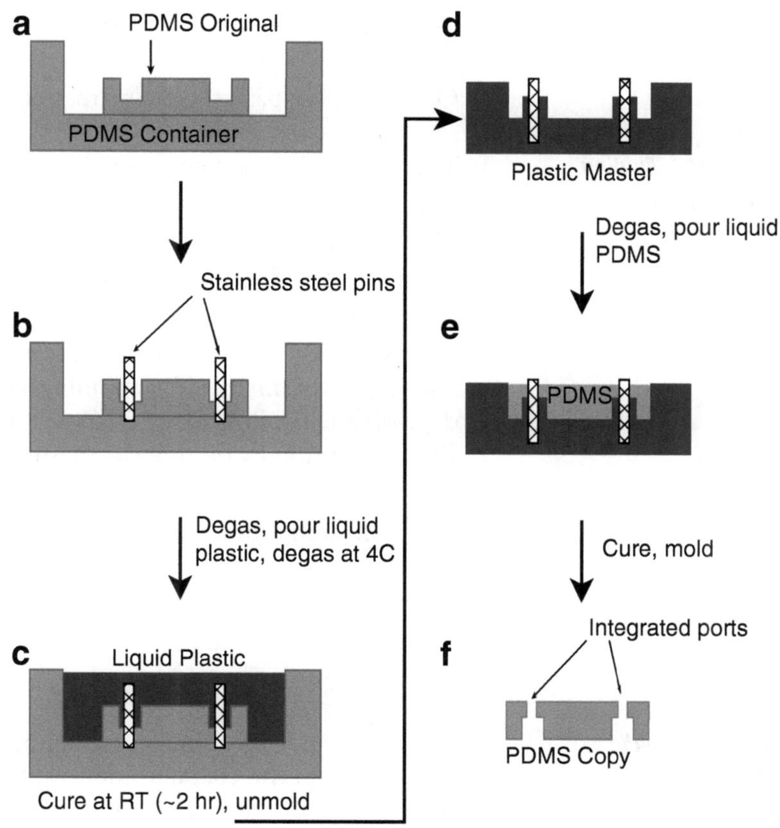

Fig. 5. Plastic master fabrication (adapted from ref. 74 reproduced by permission of The Royal Society of Chemistry). (**a**) The PDMS original is placed into a PDMS container after being punched where pins will be inserted. (**b**) Stainless steel pins are inserted in the PDMS original. (**c**) The liquid plastic (Smooth Cast 310) is best stored and prepared at 4°C. It is then poured and degas (see text) into the PDMS container and then cured at room temperature. (**d**) The stainless steel pins are transferred to the plastic master during unmolding. The plastic master can be used to cast PDMS without any further treatment. (**e**) PDMS mold are manufactured following classic procedures but cured at 65°C. (**f**) The final PDMS copy includes integrated port that are be precisely located during the process and replicated with high fidelity.

4. Prepare equal weight of Smooth Cast parts A and B. Mix parts A and B; transfer the solution to a 50-mL plastic tube and degas the mixture by spinning at $1{,}300 \times g$ for 2 min at 4°C (76).
5. Pour the polyurethane solution onto your PDMS mold (see Fig. 5c). Degas the PDMS mold/polyurethane solution for 20 min (see Note 52). Turn off the vacuum and remove the last bubble using a long fine plastic tip used to load protein gels. Cure the plastic master at room temperature for 2 h. The plastic master will gradually turn white.
6. Remove carefully the PDMS mold from the plastic master. The pins should transfer to the plastic master at that stage

(see Fig. 5d). The plastic master can be used to make PDMS molds (see Fig. 5e-f).

7. Optional: post-bake the plastic master in a 65°C oven for 4–6 h.

3.15. PDMS Molding from the Master

1. These operations are performed on a lab bench (see Note 53):
2. Use a disposable aluminum dish as a container for curing PDMS (see Note 54). Blow the master with nitrogen air and place it in the aluminum container.
3. Mix PDMS prepolymer and cross-linker at 1:10 weight ratio (see Note 55). Transfer the solution to a 50-mL plastic tube and degas the mixture by spinning at 1,300 × *g* for 2 min for degassing (76). Pour the PDMS mix on the master. Degas the PDMS solution by placing the master under vacuum in a desiccator jar until disappearance of bubbles (see Note 56).
4. Place the aluminum container on a leveled hot plate at 150°C for 10 min (77) (see Notes 57–59). Alternatively: cure the PDMS in an oven at 60°C for 2 h.
5. Carefully remove the PDMS mold from the master (see Notes 60 and 61).

3.16. Microfluidic Chip Assembly

1. Punch the ports of the microfluidic circuits using a biopsy puncher from the channel side with a diameter slightly smaller than 1/32″ for fluidic ports and slightly smaller than 1/16″ for electrode ports (see Note 62). Clean the PDMS slab with a piece of tape (Scotch 810 Magic Tape) (see Note 63) to lift off any remaining debris (see Note 64). Protect the PDMS mold with tape until use.
2. Sonicate glass slide in 2% Hellmanex III (Hellma Analytics, Germany) cleaning solution for 10 min (see Note 65).
3. Transfer the PDMS mold and glass slides to the clean room (see Note 66).
4. Just prior to use, rinse glass slides with deionized water, then with isopropanol, and dry the slides with nitrogen gas.
5. The PDMS slab is bonded to the glass slide after O_2 plasma activation of both parts (67) using a plasma cleaner. Put the PDMS slab and glass slide facing up on the tray of the plasma cleaner. Switch on the vacuum of the plasma cleaner. Fill the chamber with oxygen gas. Activate the surfaces by generating the plasma (see Notes 67 and 68).
6. Immediately after activation bring the glass slide into contact with the PDMS slab (see Note 69). Press with your fingers on the PDMS from the center outward to remove any air bubble. Act quickly as the surfaces remain activated only for a few minutes.

7. Place the bonded chip on a hotplate at 70°C for 10 min to complete the bonding process (see Note 70).

3.17. Electrodes Fabrication

1. We design electrodes as channels in the PDMS device (see Subheading 3.9) and fill them with a low melting point metal alloy (78).
2. Optional: treat the electrode channels with 3-mercaptopropyltrimethoxysilane diluted 1:10 in acetonitrile. On a 70°C hot plate, inject the solution into the electrode channel with a syringe and then immediately blow it out with nitrogen gas. This treatment increases the wetting of solder to the channels.
3. Cut a short piece of 1/16″ PEEK (polyetheretherketone) tubing (0.04″ or 1 mm inner diameter), and insert it in to the electric port that will be connected to the wire. This piece gives the port some mechanical rigidity and provides for strain relief. Strip two electric wires and insert them in the piece of PEEK tubing.
4. Set up the chip on a hot plate at 70°C for 5 min (see Note 71).
5. Load a disposable plastic syringe mounted with a blunt dispensing needle (16 gauge) with melted low-melting solder (Cerrolow-117, 47°C melting temperature) (see Note 72).
6. Inject slowly the solder into the channels while the chip is still on the hot plate. Stop the injection when the solder appears at the electrical wire, at which point release the plunger and wait for a few seconds to let the pressure equilibrate in the syringe. Remove the needle from the port and remove the chip from the hot plate when all the electrical ports are injected (see Note 73).
7. Remove the chip from the hot plate. Let cool down. Check the electrode conductivity with an electrical multimeter.
8. Affix some copper tape under the chip on the glass slide so as to mask most of the electrodes in order to confine the effects of the electric field (see Note 74).

3.18. Surface Treatment (see Note 75)

1. The protocol is adapted from ref. 79. Prepare a solution of 1H,1H,2H,2H-perfluorodecyl-trichlorosilane 1% weight in HFE 7100, store the solution in a sealed glass vial under vacuum in a desiccator jar in presence of desiccants (see Note 76).
2. Dry out the chip on a hot plate at 70°C or in an oven (see Note 77).
3. Inject the silane solution in the channels using preferably ports with online filters (see Note 78). Incubate a few minutes then blow off the solution with nitrogen gas.
4. Optional: you can repeat the process using Aquapel glass treatment (see Note 79).

3.19. Microfluidic Interconnections

1. Cell solution is flowed through 1/32″ FEP tubing with a 400-μm inner, the other solutions are flowed through 1/32″ PEEK

tubing. The interconnection between 1/32″ tubing and the chip is assured by the PDMS compliance that provides a seal around the tubing (see Note 80). This provides adequate sealing for typical fluidic pressures. The tubing is simply inserted into the ports using tweezers for better control (see Notes 81 and 82).

3.20. Setup and Experiments

1. Fluid displacement. Fluids can be displaced either by syringe pumps or pressure systems. A pressure system allows access to a wider range of droplet sizes for a specific nozzle design (80), supplies a much shorter response time, and provides a simpler way to load reagents by disregarding the presence of air bubbles in the loading vials (see Note 83). When using pressure systems, it is useful to add high resistance parts so that the outlet line has the lowest resistance in order to avoid backflow in lines and reservoirs during transitions (see Note 84).
2. Fluorescence setup. A fluorescence station for the analysis of droplets is very similar to a microscope-based flow cytometer (81, 82). It includes an inverted microscope equipped (not shown in Fig. 6) with microscope objectives (see Note 85). The excitation part of the optical train comprises at least one laser (see Fig. 6: Laser) (see Note 86) whose spot is usually shaped into a slit with cylindrical lenses (see Fig. 6: CL). The emitted fluorescence is collected by epifluorescence and transmitted to a series of photomultipliers via a series of dichroic mirrors and filters (see Fig. 6: respectively, PMT, DMi, and Fi). A series of apertures permits to block contribution of out of focus light (see Fig. 6: Ap). PMTs and fluid displacement systems are controlled by custom software developed on Labview platform.
3. Illumination. Because of the very high droplet throughput, examination of the device operation is best done with a stroboscope illumination or using a camera with a short integration time. The illumination for wide-field inspection uses a wavelength outside the range of the assay fluorescences using either some color filters or color LEDs (see Fig. 6: Illumination). This scheme permits to both monitor the operation of the device and collect fluorescence signals at the same time (see Note 87).
4. High-speed videos. High speed videos are best taken with very high-speed cameras (up to 40,000 frames per second). The steep price of these instruments is however an obstacle for most laboratories. A cheaper alternative, although not as satisfying, is to use the consumer grade high-speed EX-F1 camera from Casio that has a frame-rate of 1,200 images per second. In this case, flow-rates often need to be adjusted to a lower range than the one used for typical operation. This configuration is still

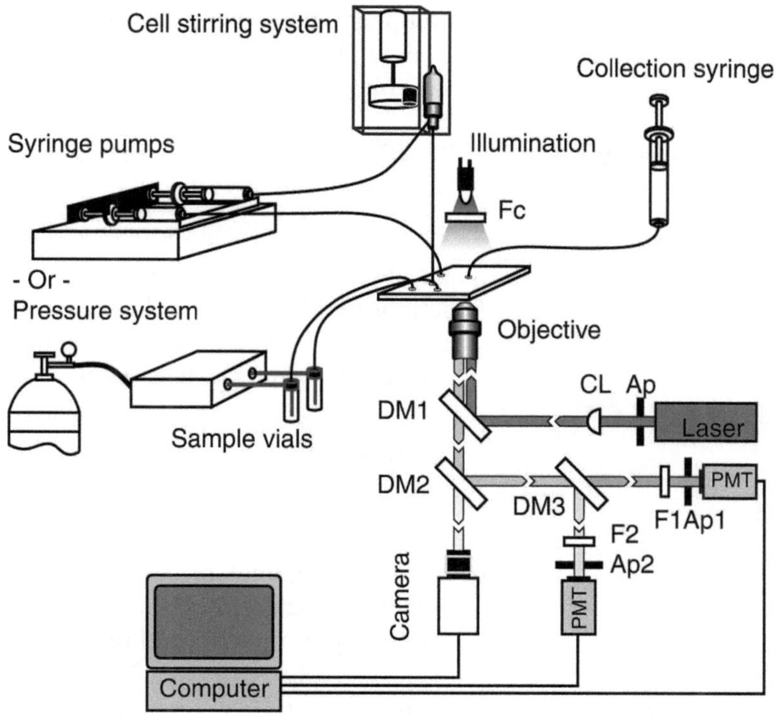

Fig. 6. The experimental setup permits to generate and manipulate droplets, and to interrogate their fluorescence. It is equipped with a fluid displacement apparatus (syringe pumps or pressure system), a cell stirring module, an illumination for wide-field inspection of device operation, a collection syringe, and an optical train that allows for fluorescent excitation and collection and wide-field inspection. The system is controlled by a custom software developed under Labview. (The module delivering the high-voltage high-frequency electric field for merging droplets is not depicted).

useful for analyzing droplet behavior and for illustration purpose. For taking videos with white and dark droplets, use PBS buffer for the white droplets and a solution of 1% wt/wt bromophenol blue in water for the dark droplets. Increase the contrast between droplets by limiting the spectral range of the illumination to the maximum absorbance of bromophenol blue (i.e., 590 nm) with a colored filter (38).

5. Measuring droplet size. Back-calculate the droplet size using the flow-rate of the aqueous phase and the droplet frequency that you can measure with the station by encapsulating a soluble fluorescent dye in the droplets at low concentration (see Note 88): $\text{roplet size} = \dfrac{\text{aqueous flow} - \text{rate}}{\text{droplet frequency}}$.
6. Electric module for droplet fusion. The high-voltage high-frequency electric field (~300 V cm^{-1}, 100 kHz) required to merge droplets is delivered by a Cold Cathode Fluorescent

Lamp inverter alimented by a DC power supply (see Note 89). The amplitude of the electric field is controlled by adjusting the voltage of the power supply (see Note 90).

3.21. Cell Injection

1. To efficiently inject cells into the microfluidic device, follow these rules:
2. Use a delivery tubing as short and as straight as possible.
3. Use a tubing material as inert as possible to minimize the interactions of cells with its walls (FEP tubing is preferable to PEEK tubing for instance), and with the biggest inner diameter possible to minimize shear stress.
4. Follow typical flow cytometry protocols: Use DNAse to digest the sticky DNA shed by dying cells; and filter cell suspension through a 40-μm cell strainer.
5. Use a stirring system for large volume of cells.
6. For small volumes of cells: fill the injection tubing with oil without surfactant, aspirate cells in the injection tubing, connect the tubing to the chip and inject cells.

3.22. Cell Stirring System

1. Different custom systems have been developed to keep cells from settling down (6, 83), and recently a commercial solution has been offered by Cetoni (Germany) (see Note 91).
2. The cell stirring system described in (6) comprises a DC motor mounted with a rotor that incorporates a rare earth magnet (see Fig. 6: Cell stirring system). The vessel consists of a vial with conical tip, a magnetic stirrer that freely rotated around a central tubing used to eject cells to the chip and an additional lateral tubing to inject fluorinated oil. The cap of the vial is modified to incorporate a modified nanoport from Idex. The system is used downward as depicted in the picture to avoid any curvature of the tube carrying the cells out to the chip. The vial is preloaded with the cell solution. Upon injection of fluorinated oil which is more dense than oil, the cell solution is displaced upward and deflected back into the central tubing (see Note 92).
3. To assemble the vessel and stirrer: Machine the base of the nanoport so it can fit into the vial cap (dotted line in Fig. 7a). Drill a 1/32″ hole at an angle in the side of the nanoport, and go through the base of the nanoport (arrow in see Fig. 7b).
4. Widen the opening of the vial cap with a 7-mm drill bit by hand (dotted line in Fig. 7c). Insert the nanoport into the cap after unscrewing the fitting. Insert a 1/32″ PEEK tubing in the side port and glue (see Fig. 7d).
5. Prepare the magnetic stirrer (see Fig. 7e): (a) Cut a piece of 1/16″ OD × 1 mm ID PEEK tubing. (b) Align and glue sequentially 2 magnets directly on the PEEK tubing (see Note 93).

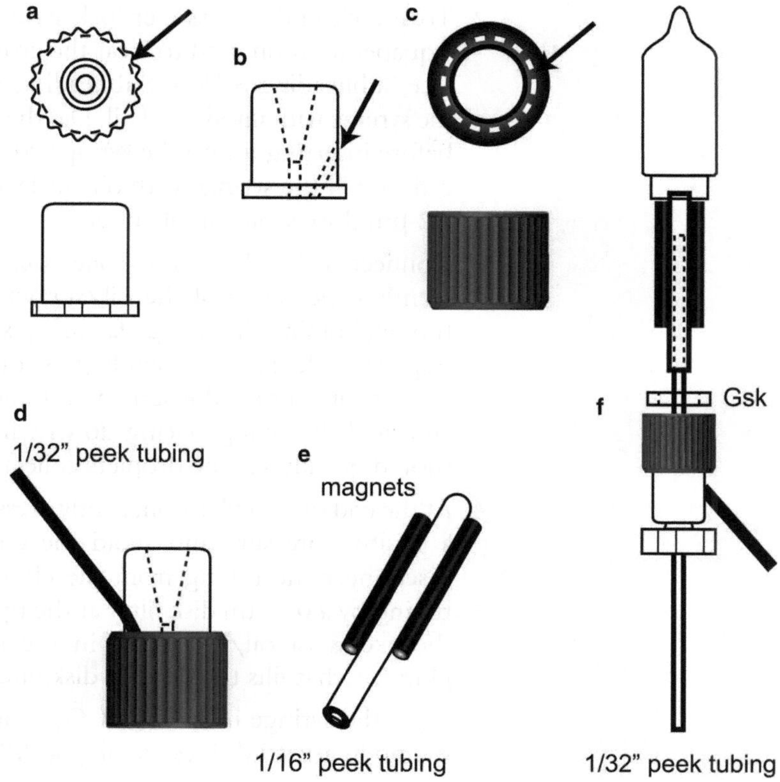

Fig. 7. Vial assembly for cell stirring system. (**a**, **b**): Modification of the nanoport which is first machined to reduce the diameter of the collar (*dotted line* in (**a**)), and then drilled at an angle to glue a 1/32″ PEEK tubing. (**c**) The vial cap is machined to adapt to the diameter of the nanoport. (**d**) The nanoport is inserted through the vial cap, and the side PEEK tubing is glued to the nanoport. (**e**) The magnet stirrer is fabricated by gluing two strong magnets to a piece of 1/16″ PEEK tubing with a large inner diameter (1 mm). (**f**) The stirring vial is assembled after inserting a central tubing into the nanoport, a gasket (Gsk) is used to assure proper sealing of the system.

6. Insert a central 1/32″ FEP tubing in the nanoport (see Fig. 7f). Insert the stirrer on the central FEP tubing. The cap is screwed on the vial; the sealing is assured by a custom rubber gasket (see Fig. 7f: Gsk).
7. When ready, open the vial and transfer cells by pipetting, close the vial and set it up next to the magnetic rotor, switch the motor on. Connect the side tubing to the oil (without surfactant) supply. Fill the vial until the air is totally pushed out of the vial and cells start flowing at the tip of the tubing. Connect the tubing to your device and start the experiment.

3.23. Cell Collection and Incubation

1. Droplets can be collected directly into a glass syringe (see Note 94), and hence ready to be reinjected with a syringe pump without transfer to another vessel (see Note 95):

2. Treat a clean dry glass luer-lock syringe with the silane or the Aquapel solution used to treat the channels using a foam swab (see Subheading 3.18 in Subheading 3.12). Thoroughly rinse the syringe with fluorinated oil. Dip the plunger in oil/surfactant before inserting it into the syringe barrel to assure good lubrication. Fill the syringe with oil/surfactant (about 1 mL) using a 2-μm disposable sample filter.
3. Connect a 1/32″ PEEK collection tubing to the syringe. Gently expel some of the oil/surfactant to fill the collection tubing, holding the syringe facing up so that no air bubble gets trapped in the syringe. Gently press on the plunger to create a positive pressure and insert the tubing at the collection outlet and held the syringe facing down when ready to collect, you should quickly see the droplets collecting in the syringe.
4. At the end of the collection, gently press on the plunger to create a positive pressure and avoid the collection of air bubbles; disconnect the tubing from the chip. Replace the collection tubing by a 0.2-μm disk filter at the tip of the syringe. Remove the excess of oil/surfactant in the syringe by pressing the plunger (that fills the 0.2-μm disk filter with oil/surfactant).
5. Store the syringe in a 37°C, CO_2 incubator facing down with the tip in a Petri dish containing a PBS buffer solution.

3.24. General Experimental Tips

1. Filter solution whenever possible: (a) Filter oil stocks through 0.2-μm filters. (b) Filter oil and aqueous phases when loading into syringe with a 2-μm disposable sample filter. (c) Filter cell solution through a cell strainer. (d) When you have recurrent problems with contaminants you can add purge-lines in your design (see step 9 in the Subheading 3.9 and Note 31) (see Note 96).
2. When connecting tubing to the chip ports: (a) Blow the tip of the tubing to remove fiber contaminants with a can of compressed air. (b) Use tweezers to carefully control the insertion of tubing into the chip ports.
3. Connections: (a) Use the appropriate fittings (see list of usual fittings and their use in the Subheading 2). (b) Properly tighten fittings (use specific tools when appropriate) (see Note 97). (c) Avoid using fittings with integrated filters as they can be quite problematic.
4. Be careful not to wear gloves when manipulating vessels containing emulsions. The static electricity conveyed by the gloves can destabilize emulsions in some instances.
5. Properly collect, label, and discard fluorinated wastes (see Note 98). Wear gloves when manipulating fluorinated oils and surfactants (see Notes 99–100): fluorinated surfactants have been shown to bioaccumulate in rat and human tissues (84, 85).

4. Notes

1. All tubings are available from Idex (USA) but from other sources too.
2. In addition to Idex and Vici catalogs, an excellent resource to learn about fittings is the "All about fittings" guide written by John Batts IV, freely available from the Idex Web site (https://www.idex-hs.com/downloads/allaboutfittings.pdf).
3. This creates a permanent fitting on the tubing, the ferrule cannot usually be reused.
4. Do not use this part to connect two 1/32″ tubings together.
5. The closest would be a dilution step by merging with a bigger droplet.
6. The Molecular Probes® Handbook—11th Edition is a very good source for finding appropriate staining dyes (http://www.invitrogen.com/site/us/en/home/References/Molecular-Probes-The-Handbook.html).
7. We are limiting the discussion to the couple of FC-3283 and HFE-7500 oils.
8. Collection and reinjection are better done using a glass syringe previously coated with Aquapel and avoiding using microtubes that may be charged with static electricity that disrupts emulsions see Subheading 3.24 in the Subheading 3.20.
9. For instance, U937 cells exhibit fried egg morphology after overnight incubation with PEG-PFPE surfactant. This morphology is believed to result from the interaction of cell membranes with the PEG head of the surfactant.
10. A nonexhaustive list of additives that can be tested: Bulk: PEG8000, PEG600, methyl cellulose of different viscosities; bulk surface active: BSA, Serum, Spermine, chitosan-PEG(amine and amide); cosurfactant: Pluronics F-68, Pluronics L-64, Tectronic 1307, Zonyl-FSO, Zonyl-FSN.
11. In absence of interactions, cells tend to drift to the walls of the well. So, inspect the walls if you cannot see any cells in the center of the well.
12. Hand-shaked emulsions do not always properly reflect the stability properties of emulsions generated on chip.
13. There is no need to use the droplet live-dead assay as a first approximation.
14. If you disturb the interface or wait too long once the emulsion is broken, you may lose a lot of cells stuck at the interface.

15. Avoid pipetting the oil phase, oil droplets tend to complicate the analysis.
16. Not including partitioning due to molecule solubility in fluorinated oils, which should be minimal in the case of fluorinated oils.
17. To test and optimize your assay signal on your station you can coflow separately your positive and negative controls with your stain just prior encapsulation. The coflow can be done either on-chip or off-chip with a T-junction from Idex to avoid designing a chip specifically for the test.
18. You can omit using cells for the leakage test when you use a colored dye. In this case, you can usually check the dye leaking into the oil phase upon simple eye inspection or using a spectrofluorimeter.
19. You can use the same approach to optimize the incubation time of the staining.
20. Instruction are for AutoCAD or AutoCAD LT only.
21. LinkCAD software can be used to repair all broken polylines.
22. The type of flow actuation is important when designing an encapsulation nozzle. A pressure-driven system is more forgiving in terms of nozzle dimensions, and permits access to a much larger range of droplet sizes compared to a flow-controlled system (80).
23. Merging can also be achieved with a laser (11). A recent approach, called picoinjection, permits the systematic and reliable addition of solution to droplets (16).
24. When both trains of droplets are generated on the same chip, they can be synchronized for perfect pairing by coupling the nozzles (86–88).
25. When designing a merging circuit it takes a few trial and errors to optimize the volume fraction of each nozzle so that the droplet pairing is optimal. The quality of the droplet pairing dictates the overall efficiency of the merging.
26. The merging box should not be so big that two droplet pairs would be inside at the same time.
27. The fabrication of incubation lines is better done by multilayer lithography (see Subheading 3.12).
28. For a given design the incubation time can be modulated by adding an oil extractor that reduces the amount of oil flowing downstream, thus reducing the overall droplet velocities. The oil extractor is a symmetrical structure composed of 2 channels perpendicular to the main channel and a series of posts to

avoid the extraction of droplets (18, 89). The oil extractor requires an extra fluid control for optimal performance.

29. This design has obvious limitations when trying to reinject droplets whose dimensions are comparable to the cell size.
30. Large droplets forced through small constrictions tend to split in smaller droplets. So, the design has to be carefully tuned if the droplets are to be further used intact.
31. If the presence of filters in the design is not possible, a lot of time and frustration can be saved by adding purge-lines that are plugged during normal operations but used to remove contaminants when clogging occurs. The purge-lines should be located across structures of the smallest dimensions.
32. Keep in mind that the resolution depends on the layer thickness.
33. For instance, when designing a long winding incubation line you should leave enough space between each turn so that the bonding could sustain reasonable pressure inside the channel and accommodate the resolution of the lithography as well.
34. Practically, it is possible to design and test a series of simple devices in about a week.
35. Once the resolution of your fabrication process is known, it should be taken into account during the designing of the circuit (see Design general rules in the Subheading 3.9).
36. Silicon wafers with (100) orientation are less fragile than wafers with (111) orientation.
37. The silicon wafers can be cleaned with a piranha wet etch (using H_2SO_4& H_2O_2) followed by a deionized water rinse. However, we usually skip this part without incidence on SU8 adhesion.
38. The layer thickness is a function of both the viscosity of the photoresist and the spin-coating conditions. The viscosity of the photoresist is very sensitive to temperature, and temperature control of the photoresist is the most important parameter to obtain reproducible layer thicknesses.
39. Start setting the wafer on the spin-coater as centered as possible.
40. A leveled hotplate with good thermal control and uniformity is recommended for use during the different baking steps.
41. A good contact between the mask and the wafer is critical to obtain good resolution. Some irisation circles should appear after the contact is made. Failure to do so would result in an air gap between the two parts and would lead to poor resolution.

42. The exposure is proportional to the lamp flux that needs to be measured regularly to compensate for the decaying performance of the illumination over time. The collimation of the lamp is also important to obtain the best resolution possible.
43. During the postbake, the circuit structures should appear gradually when observed by eye. The structures will fail to appear if the exposure is too short; conversely the structures will appear before the postbake if the exposure is too long.
44. Alternatively: the wafer can be conveniently placed on the spin-coater and the SU8 developer dispensed from a squeeze bottle. Regularly spin the wafer to remove the developed photoresist.
45. The beaker can be immersed in an ultrasonic bath to optimize development.
46. Some white traces will appear if the development is not complete. In this case dry the wafer and immerse it again in SU8 developer.
47. When using an aligner without an integrated inspection system, you can use a nonstereo monocular microscope with appropriate zoom and lighting.
48. It is difficult to add a layer of photoresist on top of a previous layer that is deep and includes some winding channels like an incubation line. In these cases, some air bubbles are often trapped during the spin-coating in the photoresist and expand during the soft-baking step.
49. The shape of the transition depends on the viscosity of the photoresist: the more viscous the photoresit the more gradual the transition (SU8-100 is more viscous than SU8-50 and so on).
50. The desiccator jar must be located inside a chemical fume hood owing to the corrosive nature of TFOCS vapors. It should not contain desiccants.
51. Without silanisation, a master would still be able to generate more than 20 PDMS molds when handled properly. A good mixing of the two pre-PDMS parts is important to extend the lifetime of a silicon-SU8 master.
52. See Note 43.
53. Ideally at 4°C.
54. Change gloves often as the PDMS solution is fairly sticky.
55. Alternatively, press a piece of aluminum foil to the bottom of a Petri dish and mold it around the edges.

56. Mix the two components thoroughly as poor mixing will result in an unevenly cured PDMS difficult to release from the master.
57. Bubbles will appear, rise to the surface, and pop. The vacuum is turned off and on as soon bubbles reach the surface. Repeat it several times to speed up the process and until all bubbles are gone.
58. Masters made of urethane plastic (Smooth Cast) do not withstand that temperature, and should be used at a maximum temperature of 60°C.
59. Overcuring will result in brittle PDMS difficult to punch cleanly and with somewhat decreased bonding properties.
60. Undercuring the PDMS will result in a very flexible slab and the diameter of the punched holes (see Subheading 2.4 in the Subheading 3.12) will be much smaller than the nominal puncher ID. Too small holes are likely to be fractured by the insertion of the 1/32″ PEEK tubing and to result in leaking ports.
61. Be very careful when removing the PDMS slab, the wafer is fragile and does not tolerate any bending.
62. It is useful to cut the PDMS slab using an angle to indicate the side with the channels.
63. The punched holes should be straight and perpendicular to the chips surface. Do not twist the puncher inside the port while punching or removing the puncher. Wipe the tip of the punch with an ethanol soaked tissue before each punch.
64. Do not use any kind of tape; lab tapes, duct tapes or electrical tapes are not appropriate.
65. Do not hesitate to press hard when rolling the tape on the channels.
66. Glass slides need to be very clean. Clean-room grade slides can be purchased from Schott. Alternatively, others reported the use of a piranha wet etching solution, or acetone bath, or Sparkleen solution to clean glass slides.
67. The key to good bonding is to use clean surfaces, wear gloves during the whole process.
68. You need to empirically figure out the parameters (oxygen pressure, flow rate, and plasma power and treatment time) which are specific to your setting.
69. Avoid over-activation: the PDMS surface would become brittle (when you can see some cracks and irisation on the surface while looking at the slab with an inclination) and unable to bond properly.

70. If the tray of the plasma cleaner is made of glass, first transfer the activated PDMS slab and glass slide onto a piece of aluminum foil.
71. Alternatively, the glass slide and the PDMS slab can be put back in the oven for 1 h at 65°C after being sandwiched between two aluminum plates and clamped using binder clips.
72. You may damage or even break the chip if the temperature of the hot plate is too high.
73. The stock of solder can be conveniently stored in a glass beaker in a 70°C oven. Once loaded, the plastic syringe can be set apart in a heating glue gun to keep the solder liquid.
74. To limit the "splashing" of melted solder on the chip and in particular into the microfluidic ports, place some low adhesion scotch tape on the different ports and around the electrical ports to limit the spreading of the solder. The solder does not wet on the tape and can be more easily removed than on the naked PDMS.
75. Alternatively, a glass slide coated with indium Tin Oxide (ITO) (Delta Technologies Ltd, Stillwater, MN) can be used to make the chips and shield the electric field. Check that the glass side does not appear speckled, if so discard the glass slide. Determine which side is coated with ITO before bonding by measuring the resistance between two points on the glass surface using a multimeter (PDMS will not bond to the ITO coated side).
76. The recent use of fluorinated polymer to make microfluidic chips alleviates the need for surface treatment (90). In that case, the chip fabrication is substantially different.
77. You can also fill the vial with Argon gas and store it at 4°C. Then take aliquots under Argon atmosphere.
78. The trichlorosilane moiety is very reactive with moisture and would precipitate and clog moist channels upon injection.
79. Always check for the presence of particles in the silane solution before use. They indicate that the solution is too old and that it would clog the channels if used.
80. Follow the same precautions as for the silane solution: Aquapel reacts strongly with water. You do not need to dilute the Aquapel solution in a fluorinated oil, inject directly into the channels. The preferred format is the applicator: be careful when cutting out the pad, removing the glass vial from the applicator, and when opening the vial with a diamond cutter. Transfer the solution into a glass syringe mounted with a valve. Simply close the valve when done after removing any air bubble in the syringe.

81. Alternative: you can glue some PEEK tubing to the ports. Plasma treat the chips with the highest settings, the chip surface should appear brittle. Insert the pieces of PEEK tubing in the ports. Put a drop of UV-curable glue (Loctite 3526 light cure adhesive) at the contact between the tubing and the PDMS chip. Cure the glue by illuminating from the glass side. Make sure to use the appropriate wavelength.
82. Make sure to cleanly cut the tubing by either using a sharp razor blade or a polymer tubing cutter (Upchurch). When cutting tubing with thin walls, double check that the tip of the tubing did not collapse.
83. When inserting the tubing in the ports, blow the tip of the tubing using a can of pressured air to remove any dust present there. The tubing is connected to syringes or vials via fittings from Idex and Vici (see Subheading 2 for more details).
84. Samples are simply loaded in vials with specially designed caps, and there is no need to push out air bubbles contrary to syringe-pump systems.
85. That can be done by inserting some pieces of PEEK tubing with small inner diameter. In this case you need to regularly change them as they tend to clog.
86. In case of high numerical aperture, you need to use coverslip instead of glass slide in the manufacturing of chips to be compatible with the short working distance of these types of objectives.
87. A precious resource describing optical trains used in flow cytometer is the book "Practical Flow Cytometry" by Howard Shapiro (91).
88. When it is not possible to develop this strategy (when all the wavelengths are used for fluorescence detection), be careful to disconnect PMTs when the illumination is switched on.
89. Use low concentrations, as some dyes can change the surface tension and hence the droplet size at high concentration.
90. See Subheading 2.
91. Do not forget to affix a conductive tape below the chip to shield the electric field and confine its effects to the merging module (see Subheading 3.17 in the Subheading 3.12).
92. http://www.cetoni.de/englisch/products/stirring_systems_nemix.html.
93. Use a FEP tubing with a 400 μm diameter as central tubing.
94. Magnets need to be glued quickly as they tend to clamp together. That can be done using UV curable glue.

95. Assure that the plunger slides smoothly and nicely without resistance; otherwise, you will create a back-pressure too high for the good operation of the device. You may want to use sand paper to adjust the sliding, be careful not to overdo or it would create leaks. Mark plunger and barrel pairs as they are adjusted by pairs and cannot be swapped.
96. When using a pressure system, you can collect droplets in a vessel similar to the one used for stirring cells without using the stirrer.
97. Use water or PBS buffer and not oil to efficiently flush out contaminants.
98. The main obvious failure mode for droplet experiments is the presence of small leaks in the fittings.
99. A specific waste stream must be dedicated to fluorinated wastes. They must not be mixed with organic or acid/base wastes.
100. Use your best judgment to comply with both this rule and the former one.

Acknowledgments

I am greatly indebted to all my former colleagues from Raindance Technologies, particularly Dr Darren Link, for introducing me to droplet microfluidics. I would like to thank Professor Helmut Strey for useful discussions and Dr Phenix-Lan Quan for her great support. This research was supported by funds from The Center for Biotechnology, an Empire State Development, Division of Science, Technology and Innovation (NYSTAR), Center for Advanced Technology and a grant from NIH-NHGRI (1 R21 HG006206-01)

References

1. Huebner A, Sharma S, Srisa-Art M et al (2008) Microdroplets: a sea of applications? Lab Chip 8:1244–1254
2. Lindstrom S, Andersson-Svahn H (2010) Overview of single-cell analyses: microdevices and applications. Lab on a Chip 10: 3363–3372
3. Song H, Chen D L, Ismagilov R F (2006) Reactions in droplets in microfluidic channels. Angew Chem Int Ed Engl 45:7336–7356
4. Teh S Y, Lin R, Hung L H et al (2008) Droplet microfluidics. Lab Chip 8:198–220
5. Taly V, Kelly B T, Griffiths A D (2007) Droplets as microreactors for high-throughput biology. ChemBioChem 8:263–272
6. Brouzes E, Medkova M, Savenelli N et al (2009) Droplet microfluidic technology for single-cell high-throughput screening. Proc Natl Acad Sci USA 106:14195–14200
7. Clausell-Tormos J, Lieber D, Baret J C et al (2008) Droplet-based microfluidic platforms for the encapsulation and screening of Mammalian cells and multicellular organisms. Chem Biol 15:427–437
8. Holtze C, Rowat A C, Agresti J J et al (2008) Biocompatible surfactants for water-in-fluorocarbon emulsions. Lab Chip 8:1632–1639
9. Koster S, Angile F E, Duan H et al (2008) Drop-based microfluidic devices for encapsulation of single cells. Lab Chip 8:1110–1115

10. Ahn K, Agresti J, Chong H et al (2006) Electrocoalescence of drops synchronized by size-dependent flow in microfluidic channels. Applied Physics Letters 88:264105
11. Baroud C N, de Saint Vincent M R, Delville J P (2007) An optical toolbox for total control of droplet microfluidics. Lab Chip 7:1029–1033
12. Chabert M, Dorfman K D, Viovy J L (2005) Droplet fusion by alternating current (AC) field electrocoalescence in microchannels. Electrophoresis 26:3706–3715
13. Link D R, Grasland-Mongrain E, Duri A et al (2006) Electric control of droplets in microfluidic devices. Angew Chem Int Ed Engl 45:2556–2560
14. Niu X, Gulati S, Edel J B et al (2008) Pillar-induced droplet merging in microfluidic circuits. Lab Chip 8:1837–1841
15. Priest C, Herminghaus S, Seemann R (2006) Controlled electrocoalescence in microfluidics: Targeting a single lamella. Applied Physics Letters 89:134101
16. Abate A R, Hung T, Mary P et al (2010) High-throughput injection with microfluidics using picoinjectors. Proc Natl Acad Sci USA 107:19163–19166
17. Song H, Tice J D, Ismagilov R F (2003) A microfluidic system for controlling reaction networks in time. Angew Chem Int Ed Engl 42:768–772
18. Frenz L, Blank K, Brouzes E et al (2009) Reliable microfluidic on-chip incubation of droplets in delay-lines. Lab on a Chip 9:1344–1348
19. Ahn K, Kerbage K, Hunt T P et al (2006) Dielectrophoretic manipulation of drops for high-speed microfluidic sorting devices. Applied Physics Letters 88:024104
20. Sarrazin F, Prat L, Di Miceli N et al (2007) Mixing characterization inside microdroplets engineered on a microcoalescer. Chemical Engineering Science 62:1042–1048
21. Song H, Ismagilov R F (2003) Millisecond Kinetics on a Microfluidic Chip Using Nanoliters of Reagents. J. Am. Chem. Soc. 125:14613–14619
22. Chabert M, Viovy J L (2008) Microfluidic high-throughput encapsulation and hydrodynamic self-sorting of single cells. Proc Natl Acad Sci USA 105:3191–3196
23. Edd J F, Di Carlo D, Humphry K J et al (2008) Controlled encapsulation of single-cells into monodisperse picolitre drops. Lab Chip 8:1262–1264
24. Sgro A E, Allen P B, Chiu D T (2007) Thermoelectric manipulation of aqueous droplets in microfluidic devices. Anal Chem 79:4845–4851
25. Baret J C, Beck Y, Billas-Massobrio I et al (2010) Quantitative cell-based reporter gene assays using droplet-based microfluidics. Chem Biol 17:528–536
26. Huebner A, Srisa-Art M, Holt D et al (2007) Quantitative detection of protein expression in single cells using droplet microfluidics. Chem Commun 12:1218–1220
27. Boedicker J Q, Li L, Kline T R et al (2008) Detecting bacteria and determining their susceptibility to antibiotics by stochastic confinement in nanoliter droplets using plug-based microfluidics. Lab Chip 8:1265–1272
28. Luo C, Yang X, Fu Q et al (2006) Picoliter-volume aqueous droplets in oil: electrochemical detection and yeast cell electroporation. Electrophoresis 27:1977–1983
29. Xiao K, Zhang M, Chen S et al (2010) Electroporation of micro-droplet encapsulated HeLa cells in oil phase. ELECTROPHORESIS 31:3175–3180
30. Zhan Y, Wang J, Bao N et al (2009) Electroporation of cells in microfluidic droplets. Anal Chem 81:2027–2031
31. He M, Edgar J S, Jeffries G D et al (2005) Selective encapsulation of single cells and subcellular organelles into picoliter- and femtoliter-volume droplets. Anal Chem 77:1539–1544
32. Novak R, Zeng Y, Shuga J et al (2011) Single-cell multiplex gene detection and sequencing with microfluidically generated agarose emulsions. Angew Chem Int Ed Engl 50:390–395
33. Vijayakumar K, Gulati S, deMello A J et al (2010) Rapid cell extraction in aqueous two-phase microdroplet systems. Chemical Science 1:447–452
34. Konry T, Dominguez-Villar M, Baecher-Allan C et al (2011) Droplet-based microfluidic platforms for single T cell secretion analysis of IL-10 cytokine. Biosensors and Bioelectronics 26:2707–2710
35. Chen D, Du W, Liu Y et al (2008) The chemistrode: a droplet-based microfluidic device for stimulation and recording with high temporal, spatial, and chemical resolution. Proc Natl Acad Sci USA 105:16843–16848
36. Liu W, Kim H J, Lucchetta E M et al (2009) Isolation, incubation, and parallel functional testing and identification by FISH of rare microbial single-copy cells from multi-species mixtures using the combination of chemistrode and stochastic confinement. Lab Chip 9:2153–2162
37. Joensson H N, Samuels M L, Brouzes E R et al (2009) Detection and analysis of cell surface biomarkers expressed at extremely low levels using enzymatic amplification in microfluidic droplets. Angew Chem Int Ed Engl 48:2518–2521
38. Agresti J J, Antipov E, Abate A R et al (2010) Ultrahigh-throughput screening in drop-based

microfluidics for directed evolution. Proc Natl Acad Sci USA 107:4004–4009
39. Baret J C, Miller O J, Taly V et al (2009) Fluorescence-activated droplet sorting (FADS): efficient microfluidic cell sorting based on enzymatic activity. Lab Chip 9:1850–1858
40. Kumaresan P, Yang C J, Cronier S A et al (2008) High-throughput single copy DNA amplification and cell analysis in engineered nanoliter droplets. Anal Chem 80:3522–3529
41. Zeng Y, Novak R, Shuga J et al (2010) High-Performance Single Cell Genetic Analysis Using Microfluidic Emulsion Generator Arrays. Anal Chem 82:3183–3190
42. Johnston K P, Harrison K L, Clarke M J et al (1996) Water-in-Carbon Dioxide Microemulsions: An Environment for Hydrophiles Including Proteins. Science 271:624–626
43. Schmitz C H, Rowat A C, Koster S et al (2009) Dropspots: a picoliter array in a microfluidic device. Lab Chip 9:44–49
44. Huebner A, Olguin L F, Bratton D et al (2008) Development of Quantitative Cell-Based Enzyme Assays in Microdroplets. Analytical Chemistry 80:3890–3896
45. Shim J-u, Olguin L F, Whyte G et al (2009) Simultaneous Determination of Gene Expression and Enzymatic Activity in Individual Bacterial Cells in Microdroplet Compartments. Journal of the American Chemical Society 131:15251–15256
46. Granieri L, Baret J-C, Griffiths A D et al (2010) High-Throughput Screening of Enzymes by Retroviral Display Using Droplet-Based Microfluidics. Chemistry & biology 17:229–235
47. Srisa-Art M, Bonzani I C, Williams A et al (2009) Identification of rare progenitor cells from human periosteal tissue using droplet microfluidics. Analyst 134:2239–2245
48. Abbyad P, Tharaux P-L, Martin J-L et al (2010) Sickling of red blood cells through rapid oxygen exchange in microfluidic drops. Lab on a Chip 10:2505–2512
49. Bai Y, He X, Liu D et al (2010) A double droplet trap system for studying mass transport across a droplet-droplet interface. Lab on a Chip 10:1281–1285
50. Courtois F, Olguin L F, Whyte G et al (2009) Controlling the Retention of Small Molecules in Emulsion Microdroplets for Use in Cell-Based Assays. Analytical Chemistry 81:3008–3016
51. Wootton R C, Demello A J (2010) Microfluidics: Exploiting elephants in the room. Nature 464:839–840
52. Wu N, Courtois F, Zhu Y et al (2010) Management of the diffusion of 4-methylumbelliferone across phases in microdroplet-based systems for in vitro protein evolution. ELECTROPHORESIS 31:3121–3128
53. Kreutz J E, Shukhaev A, Du W et al (2010) Evolution of catalysts directed by genetic algorithms in a plug-based microfluidic device tested with oxidation of methane by oxygen. J Am Chem Soc 132:3128–3132
54. Tewhey R, Warner J B, Nakano M et al (2010) Microdroplet-based PCR enrichment for large-scale targeted sequencing. Nat Biotechnol 27:1025–1031
55. Holt D J, Payne R J, Abell C (2010) Synthesis of novel fluorous surfactants for microdroplet stabilisation in fluorous oil streams. Journal of Fluorine Chemistry 131:398–407
56. Holt D J, Payne R J, Chow W Y et al (2010) Fluorosurfactants for microdroplets: Interfacial tension analysis. Journal of Colloid and Interface Science 350:205–211
57. Roach L S, Song H, Ismagilov R F (2005) Controlling nonspecific protein adsorption in a plug-based microfluidic system by controlling interfacial chemistry using fluorous-phase surfactants. Anal Chem 77:785–796
58. Baroud C N, Gallaire F, Dangla R (2010) Dynamics of microfluidic droplets. Lab Chip 10:2032–2045
59. Di Carlo D, Irimia D, Tompkins R G et al (2007) Continuous inertial focusing, ordering, and separation of particles in microchannels. Proc Natl Acad Sci USA 104:18892–18897
60. Humphry K J, Kulkarni P M, Weitz D A et al (2010) Axial and lateral particle ordering in finite Reynolds number channel flows. Phys. Fluids 22:081703
61. Niu X, Gielen F, deMello A J et al (2009) Electro-Coalescence of Digitally Controlled Droplets. Analytical Chemistry 81:7321–7325
62. Abate A R, Chen C H, Agresti J J et al (2009) Beating Poisson encapsulation statistics using close-packed ordering. Lab Chip 9:2628–2631
63. Srisa-Art M, deMello A J, Edel J B (2009) High-throughput confinement and detection of single DNA molecules in aqueous microdroplets. Chemical Communications6548–6550
64. Rowat A C, Weitz D A (2008) How to easily punch holes in a PDMS microfluidic device. Lab on a Chip. Chips & Tips. http://www.rsc.org/Publishing/Journals/lc/Chips_and_Tips/punching_holes.asp.
65. Eddings M A, Johnsson M A, Gale B K (2008) Determining the optimal PDMS–PDMS bonding technique for microfluidic devices. Journal of Micromechanics and Microengineering 18:067001
66. Fuerstman M J, Lai A, Thurlow M E et al (2007) The pressure drop along rectangular

microchannels containing bubbles. Lab Chip 7:1479–1489

67. Duffy D C, McDonald J C, Schueller O J A et al (1998) Rapid Prototyping of Microfluidic Systems in Poly(dimethylsiloxane). Anal. Chem. 70:4974–4984
68. McDonald J C, Duffy D C, Anderson J R et al (2000) Fabrication of microfluidic systems in poly(dimethylsiloxane). Electrophoresis 21:27–40
69. Qin D, Xia Y, Whitesides G M (1996) Rapid prototyping of complex structures with feature sizes larger than 20 μm. Advanced Materials 8:917–919
70. Microchem. PROCESSING GUIDELINES FOR: SU-8 2025, SU-8 2035, SU-8 2050 and SU-8 2075. Data Sheet. http://www.microchem.com/products/pdf/SU-82000DataSheet82025thru82075Ver82004.pdf.
71. Hung L H, Lin R, Lee A P (2008) Rapid microfabrication of solvent-resistant biocompatible microfluidic devices. Lab Chip 8:983–987
72. Tan J L, Tien J, Pirone D M et al (2003) Cells lying on a bed of microneedles: an approach to isolate mechanical force. Proc Natl Acad Sci USA 100:1484–1489
73. Wang J, Zheng M, Wang W et al (2010) Optimal Protocol for Molding PDMS with a PDMS master. Lab on a Chip. Chips & Tips. http://www.rsc.org/Publishing/Journals/lc/Chips_and_Tips/moldingPDMS.asp.
74. Desai S P, Freeman D M, Voldman J (2009) Plastic masters-rigid templates for soft lithography. Lab Chip 9:1631–1637
75. Estevez-Torres A, Yamada A, Wang L (2009) An inexpensive and durable epoxy mold for PDMS. Lab on a Chip. Chips & Tips. http://www.rsc.org/Publishing/Journals/lc/Chips_and_Tips/epoxy_mould.asp.
76. LaFratta C N (2010) Degas PDMS in Two Minutes Using a Centrifuge. Lab on a Chip. Chips & Tips. http://www.rsc.org/Publishing/Journals/lc/Chips_and_Tips/degas_PDMS.asp.
77. O'Neil A, Soo Hoo J, Walker G (2006) Rapid curing of PDMS for microfluidic applications. Lab on a Chip. Chips & Tips. http://www.rsc.org/Publishing/Journals/lc/Chips_and_Tips/index.asp.
78. Siegel A C, Shevkoplyas S S, Weibel D B et al (2006) Cofabrication of electromagnets and microfluidic systems in poly(dimethy-lsiloxane). Angew Chem Int Ed Engl 45:6877–6882
79. Clausell-Tormos J, Griffiths A D, Merten C A (2010) An automated two-phase microfluidic system for kinetic analyses and the screening of compound libraries. Lab on a Chip 10:1302–1307
80. Ward T, Faivre M, Abkarian M et al (2005) Microfluidic flow focusing: drop size and scaling in pressure versus flow-rate-driven pumping. Electrophoresis 26:3716–3724
81. Lindmo T, Steen H B (1979) Characteristics of a simple, high-resolution flow cytometer based o a new flow configuration. Biophys J 28:33–44
82. Steen H B, Lindmo T (1979) Flow cytometry: a high-resolution instrument for everyone. Science 204:403–404
83. Baret J C (2009) A remote syringe for cells, beads and particle injection in microfluidic channels. Lab on a Chip. Chips & Tips. http://www.rsc.org/Publishing/Journals/lc/Chips_and_Tips/remote_syringe.asp.
84. Hinderliter P M, DeLorme M P, Kennedy G L (2006) Perfluorooctanoic acid: relationship between repeated inhalation exposures and plasma PFOA concentration in the rat. Toxicology 222:80–85
85. So M K, Yamashita N, Taniyasu S et al (2006) Health risks in infants associated with exposure to perfluorinated compounds in human breast milk from Zhoushan, China. Environ Sci Technol 40:2924–2929
86. Frenz L, Blouwolff J, Griffiths A D et al (2008) Microfluidic production of droplet pairs. Langmuir 24:12073–12076
87. Hashimoto M, Shevkoplyas S S, Zasonska B et al (2008) Formation of bubbles and droplets in parallel, coupled flow-focusing geometries. Small 4:1795–1805
88. Hong J, Choi M, Edel J B et al (2010) Passive self-synchronized two-droplet generation. Lab on a Chip 10:2702–2709
89. Prat L, Sarrazin F, Tasseli J et al (2006) Increasing and decreasing droplets velocity in microchannels. Microfluidics and Nanofluidics Volume 2:271–274
90. Begolo S, Colas G, Viovy J L et al (2011) New family of fluorinated polymer chips for droplet and organic solvent microfluidics. Lab Chip 11:508–512
91. Shapiro H M (2003) Practical flow cytometry. Wiley-Liss, New York

Chapter 11

Screening of Antigen-Specific Antibody-Secreting Cells

Hiroyuki Kishi, Aishun Jin, Tatsuhiko Ozawa, Kazuto Tajiri, Tsutomu Obata, and Atsushi Muraguchi

Abstract

Screening of antigen-specific antibody-producing cells is a key step for obtaining antigen-specific monoclonal antibodies. In murine system, hybridoma between B-lymphocytes and myeloma cells is used to screen and produce antigen-specific monoclonal antibodies. In human system, good hybridoma-producing system is not available. Instead, transformation of B-lymphocytes with Epstein–Barr viruses is used to obtain antibody-secreting cell lines. Furthermore, phage-display system using molecular biology is recently used to obtain antigen-specific human monoclonal antibodies. Here, we describe the new method for screening antigen-specific antibody-secreting cells at single-cell levels using microwell-array chips. The system can be applied to screen antigen-specific antibody-secreting cells from any animal species.

Key words: Antigen-specific antibody, Antibody-secreting cell, Microwell-array chip, Immunospot array assay on chip

1. Introduction

Producing antibody-secreting cell lines from B-lymphocytes is the basal technique to screen and obtain antigen-specific antibodies. In murine system, hybridoma between B-lymphocytes and myeloma cells is produced to obtain antibody-secreting cell lines (1). In other animal species including human, efficient and stable hybridoma-producing system is not available. In human system, Epstein–Barr virus-transformed B-cell lines are often used to produce cell lines that produce human monoclonal antibodies (2, 3). In addition, library of phages that express antibodies on the surface is produced using recombinant molecular biology techniques and used to screen antigen-specific human antibodies (4, 5). Time-consuming and laborious step in both methods is that of preparing libraries of cells or phages that produce antibodies. It takes a month to several months to obtain the libraries and it is not so easy for any laboratories to

Sara Lindström and Helene Andersson-Svahn (eds.), *Single-Cell Analysis: Methods and Protocols*, Methods in Molecular Biology, vol. 853, DOI 10.1007/978-1-61779-567-1_11, © Springer Science+Business Media, LLC 2012

prepare good libraries (containing cells or phages producing antibodies with large varieties of antigen specificities), which makes it difficult to produce personal libraries from individual volunteers or patients who produce antigen-specific antibodies in blood.

Recently, we have developed microwell array chips that have 45,000–230,000 of microwells whose size and shape are just fit to capture single cells in each microwell (6–8). By arraying live human B-lymphocytes on the chip, we stimulated them with antigen and analyzed the alteration of intracellular Ca^{2+} concentration in antigen-stimulated B-lymphocytes on the chip. Since intracellular Ca^{2+} concentration transiently increases by the signals from B-cell antigen receptor, i.e., the cell surface immunoglobulin on B-lymphocytes, we can identify B-lymphocytes that produce antigen-specific antibodies on the chip. We have also tried and detected antigen-specific B-lymphocytes on the chip by arraying live human B-lymphocytes on the chip and detecting the binding of fluorescence-labeled antigen to B-cell antigen receptors on antigen-specific B-lymphocytes on the chip (9). We could successfully detect antigen-specific human B-lymphocytes on the chip using these protocols. The precision of antigen-specific B-lymphocyte detection, however, was not so high for these two protocols because of the background noises in the assay. So we have recently developed the third protocol to detect lymphocytes that produce antigen-specific antibodies on the chip (10). To this end, we analyzed antigen specificity of the secreted antibodies from single lymphocytes that were accommodated in microwells on the chip. As the corresponding microwell array chips, Love et al. have developed microengraving method that uses microwell array chips with microwells of 50 μm diameter and depth (11). For analyzing single cells, they need to adjust the cell concentration by limiting dilution. So it is difficult to trap single cells in most of wells. Deutsch et al. (12) as well as Biran and Walt (13) reported the microwell arrays that could accommodate single cell in each well, whereas their microwells were not suitable to detect antibody secretion from the accommodated cells and to retrieve the detected target cells. In these contexts, our microwell array chips are unique for their capacities of capturing single cells in each microwell, detecting antibody secretion from single cells, and enabling the retrieval of objective cells.

2. Materials

2.1. Preparation of Antigen-Coated Microwell Array Chips

1. Kimwipe™ paper (Kimberly-Clark Worldwide, Inc.).
2. Lipidure: 5% Lipidure BL-103 (NOF, Tokyo, Japan). Dilute the Lipidure to 0.01% (500-fold) in phosphate-buffered saline (PBS) before use. Keep it at room temperature until use.

3. Vacuum pump.

2.2. Preparation of Antibody-Secreting Cells

1. Heparin: Store at 4°C.
2. Syringe.
3. Needle: 21 G.
4. Ficoll-Conray solution: Lymphosepal I (IBL, Fujioka, Japan).
5. PBS: Weigh 8.0 g NaCl, 0.2 g KCl, 1.44 g of Na_2HPO_4, and 0.24 g KH_2PO_4 and transfer to a 1-L glass beaker containing about 900 mL of water (ultrapure water prepared by purifying deionized and distilled water to attain a sensitivity of 18 MΩ cm at 25°C). Mix and make up to 1 L with water. Sterilize by autoclaving at 121°C for 20 min.
6. Hemocytometer.
7. Turk solution.
8. CD138 MicroBeads (Miltenyi Biotec, Bergisch Gladbach, Germany) (see Note 1).
9. AutoMACS pro (Miltenyi Biotec).
10. AutoMACS column (Miltenyi Biotec).
11. Cell culture medium: RPMI1640 (Invitrogen, Tokyo, Japan) supplemented with 10% fetal calf serum, 100 U/mL penicillin, 100 μg/mL streptomycin, and 50 μM 2-mercaptoethanol (see Note 2).

2.3. Immunospot Array Assay on a Chip

1. Microwell array chip containing 90,000 microwells (10 μm in diameter) (see Note 3).
2. PBS.
3. Antigen (see Note 4).
4. Lipidure BL-103 (provided as 5 wt% solution from NOF Corporation, Tokyo, Japan) (see Note 5).
5. PBS-L: PBS containing 0.01% Lipidure BL-103.
6. Cy3-conjugated anti-IgG; store according to the manufacturer's instruction (see Note 6).
7. Cell Trace Oregon green 488 (Invitrogen): Dissolve in dimethyl sulfoxide at 1 mg/mL and store at –20°C. Dilute to 0.5 μg/mL just prior to use with PBS-L.
8. CO_2 incubator: 5% CO_2 in the air, 37°C.
9. Incubation box (see Note 7).
10. Fluorescence microscope with halogen lamp, optical filters to observe FITC (for observing Oregon Green) and Cy3.
11. Digital camera for a microscope, which can observe Oregon Green and Cy3.

2.4. Retrieval of Antigen-Specific Antibody-Secreting Cells

1. Fluorescence microscope.
2. Micromanipulator (TransferMan NK2, Eppendorf, Hamburg, Germany).
3. Microinjector (CellTram vario, Eppendorf).
4. Capillary (12 μm in diameter, Primetech, Tokyo, Japan).
5. Paraffin oil.

3. Methods

3.1. Preparation of Antigen-Coated Microwell Array Chips

Antigen is coated onto the chip overnight, and then the chip is used for immunospot array assay on a chip (ISAAC).

1. Add 100 μL of antigen (see Notes 4 and 8) in PBS onto the well area on a microwell array chip.
2. Put the chip in a humidified black box.
3. Incubate the chip overnight at 4°C or room temperature (see Note 9).
4. Wash the chip by aspirating the antigen solution with a Kimwipe paper (or a filter paper) and then adding 100 μL of PBS-L.
5. Repeat step 3.
6. Remove the air in microwells using a vacuum pump.
7. Incubate the chip for 15 min at room temperature.
8. Wash the chip as in step 3 using cell culture medium instead of PBS-L (see Note 10).
9. Use the chip for ISAAC.

3.2. Preparation of Antibody-Secreting Cells

Human experiments should be performed with the approval of the Ethical Committee at the belonging organization, and informed consent should be obtained from the subjects.

1. Collect 50 mL of peripheral blood into syringe from vaccinated volunteer at 7th day post vaccination (see Note 11).
2. Divide 5 mL of blood into 15-mL conical tubes that contain 5 mL of PBS.
3. Insert a sterile Pasteur pipette into blood (Fig. 1 and see Note 12).
4. Add 3 mL of Lymphosepal I to the tube through the Pasteur pipette. Lymphosepal I goes under the blood.
5. Centrifuge the tube at 740 ×g (2,000 rpm) for 30 min at room temperature (see Note 13).

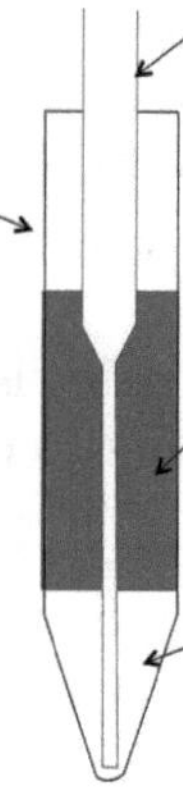

Fig. 1. Layering peripheral blood over Lymphosepal I. Sterile Pasteur pipette is inserted into twofold diluted peripheral blood. Lymphosepal I is, then, poured into the blood from top of the Pasteur pipette.

6. Mononuclear cells are layered over Lymphosepal I. Harvest the mononuclear cells using a Pasteur pipette (see Note 14) and transfer them to another 15-mL conical tube.
7. Dilute cells with PBS to 10 mL and centrifuge at 740 × *g* at room temperature.
8. Discard the supernatant with decantation, resuspend cells with approximately 10 mL of PBS, and centrifuge cells at 420 × *g* (1,500 rpm) for 5 min at room temperature.
9. Repeat step 9.
10. Discard the supernatant with decantation, and resuspend cells with 10 mL of PBS.
11. Count the cell number using Turk solution (see Note 15).
12. Centrifuge cells at 420 × *g* for 5 min at room temperature.
13. Remove the supernatant completely and add 90 μL of PBS per 10^7 cells.
14. Add 10 μL of CD138-microbeads to the cells.
15. Incubate the cells at 4°C for 15 min.
16. Add 1–2 mL of PBS to the cells and centrifuge them at 420 × *g* for 5 min at room temperature.
17. Remove the supernatant as completely as possible, and add 500 μL of PBS.
18. Separate $CD138^+$ cells using AutoMACS Pro according to the manufacturer's instruction.
19. Count the cell number (see Note 16).

20. Spin down cells at $420 \times g$ for 5 min at room temperature. Discard the supernatant and resuspend cells in cell culture medium to the cell density of 1.8×10^6 cells/mL.
21. Keep cells at room temperature until use.

3.3. Immunospot Array Assay on a Chip (See Note 17)

When the chip is incubated, put the chip in a humidified black box in order not to dry the chip.

1. Remove the medium by aspirating it with a Kimwipe paper (or a filter paper), and then add 100 μL of cells (1.8×10^5 cells).
2. Incubate the cells at room temperature for 5 min until the cells settle down into the wells.
3. Mix the cell suspension with a pipette for making cells outside of wells floating and incubate the chip for 5 min at room temperature (see Note 18).
4. Repeat step 3 once more.
5. Mix the cell suspension with a pipette for making cells outside of wells floating and remove the cell suspension by aspirating it with a Kimwipe paper (or a filter paper). Add 100 μL of cell culture medium.
6. Repeat step 5 twice more.
7. Incubate cells at 37°C for 3 h in a humidified CO_2 incubator.
8. Remove the medium by aspirating it with a Kimwipe paper (or a filter paper), and then add 100 μL of PBS.
9. Repeat step 8 twice more.
10. Remove PBS by aspirating it with a Kimwipe paper (or a filter paper), then add 100 μL of Cy3-conjugated anti-IgG (1 μg/mL in PBS-L), and incubate the chip at room temperature for 30 min.
11. Remove antibody solution with a Kimwipe paper (or a filter paper), and then add 100 μL of PBS.
12. Repeat step 11 twice more.
13. Remove PBS by aspirating it with a Kimwipe paper (or a filter paper), then add 100 μL of Cell Trace Oregon green 488 (0.5 μg/mL in PBS-L), and incubate at room temperature for 5 min (see Note 19).
14. Wash the chip as in step 11 three times.
15. Observe antibody secretion with Cy3 signals (see Note 20) and cells with Oregon green signals under a fluorescence microscope (see Fig. 2).

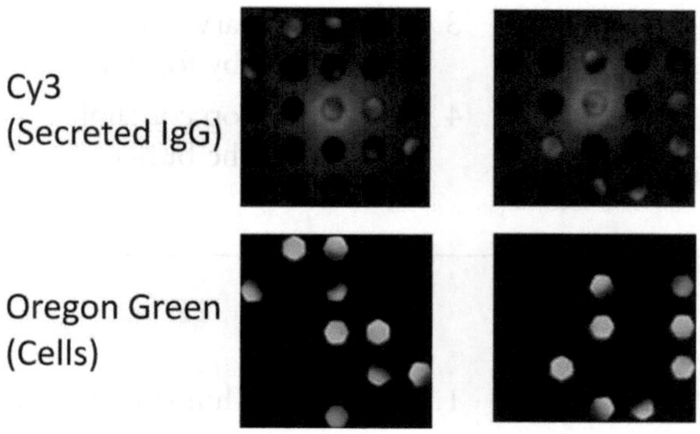

Fig. 2. ISAAC signals. *Top*, Cy3 signals showing secreted IgG. *Bottom*, Oregon green signals showing cells in microwells.

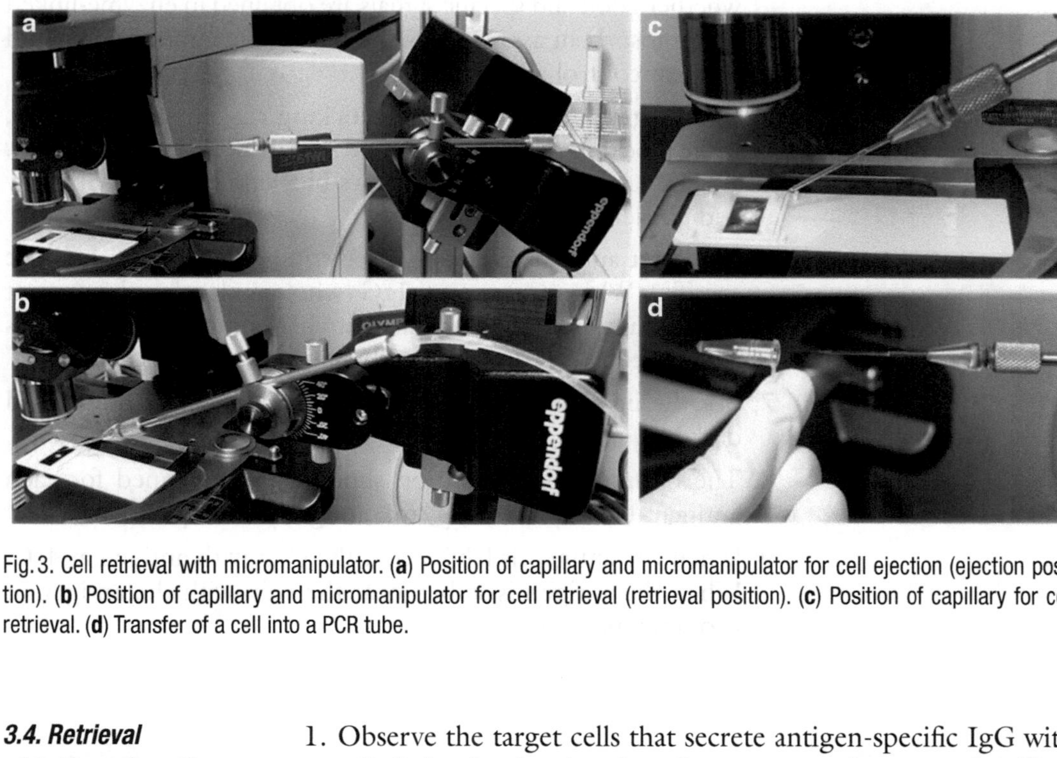

Fig. 3. Cell retrieval with micromanipulator. (**a**) Position of capillary and micromanipulator for cell ejection (ejection position). (**b**) Position of capillary and micromanipulator for cell retrieval (retrieval position). (**c**) Position of capillary for cell retrieval. (**d**) Transfer of a cell into a PCR tube.

3.4. Retrieval of Antigen-Specific Antibody-Secreting Cells

1. Observe the target cells that secrete antigen-specific IgG with a Cy3 circular signal under a fluorescence microscope (capillary and micromanipulator is at "ejection position" as in Fig. 3a) (see Note 21).
2. Bring the capillary and the micromanipulator to "retrieval position," as in Fig. 3b. Adjust the tip of the capillary to the target-cell position under a microscope using a CellTransferman NK2 (Fig. 3c). Aspirate cells with a microaspirator (CellTram Vario).

3. Bring a capillary to the "ejection position," and eject cells from the capillary by forming a tiny drop (~0.2 μL) of buffer.
4. Transfer a drop containing a cell to RT-PCR buffer by adding the drop to the buffer (Fig. 3d, see Note 22).

4. Notes

1. For isolating human CD138-positive plasma cells.
2. 2-mercaptoethanol is optional.
3. The right of chip is wholly owned by Vivalis (Nantes, France).
4. Antigen should be dissolved in PBS in the absence of stabilizer, such as bovine serum albumin. It is recommended to check whether good and specific signals are obtained in enzyme-linked immunosorbent assay (ELISA) by using the same antigen- and positive-control antiserum.
5. Blocking reagent: Any other blocking reagents can be used, if the blocking reagents give good signal/noise ratio in ELISA. We use Lipidure because it shows blocking effect very rapidly and efficiently.
6. To detect antigen-specific IgG-secreting cells, use anti-IgG (Fc specific). When you use anti-IgG (whole molecule), the reagent contains anti-immunoglobulin light-chain antibodies and it reacts with other class of antibody, including IgM, IgA, and IgE.
7. For the incubation of chip, the chip was put in a humidified dark box as shown in Fig. 4.
8. The optimum concentration should be determined for each antigen. Usually, 1–10 μg/mL antigen is enough.
9. When the antigen is labile, coat the chip with antigen at 4°C. When the antigen is stable, coat the chip with the antigen at room temperature.

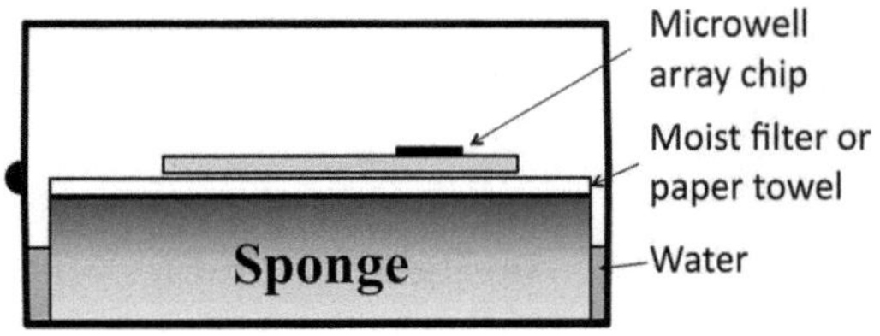

Fig. 4. Humidified dark box for the incubation of the microwell array chip. To avoid drying out of the solution on the chip during the incubation, microwell array chip was incubated in the humidified box. Sponge was humidified with water and a wet paper towel or filter paper was put on it. The chip was put on the filter paper for the incubation.

10. Be careful not to dry the chip. When aspirating the buffer, add the medium as quickly as possible.
11. The number of antigen-specific antibody-secreting cells that circulate in peripheral blood peaks 7 days after the vaccination, and then rapidly decreases within several days (14).
12. Glass Pasteur pipettes are sterilized at 180°C for 2 h in a dry incubator.
13. Brake should be off.
14. You can first remove the upper layer that contains plasma and most of platelets and then harvest the mononuclear cells.
15. Mix cells well with a Pasteur pipette. Harvest 10 μL of cell suspension and mix with 90 μL of Turk solution. Mix cell suspension well, transfer an aliquot of cell suspension to a hemocytometer, and count the cell number.
16. Volume of CD138-positive cells is approximately 2 mL after separation by AutoMACS Pro.
17. The ISAAC is covered by patents that have been exclusively licensed to Vivalis (Nantes, France).
18. The cells inside of wells tend to stay in the wells during the mixing of cell suspension. Mix cells mildly but strongly enough to float cells outside of wells.
19. Oregon Green fluorescence is strong and can be detected with optical filter for Cy3. Too strong Oregon Green fluorescence interfered for the observation of Cy3 signals. Adjust the concentration of Oregon Green and incubation time to label cells with optimal strength of the fluorescence signals.
20. Antibodies secreted from a cell that is in a microwell are diffused out to all directions. The bound secreted antibody on the chip is observed as a circular signal of Cy3. It looks like a doughnut and is easily distinguished from noise signals.
21. Be careful not to puncture fingers or eyes with the glass capillary. Wear glasses to protect eyes.
22. Be careful not to puncture fingers or eyes with a capillary. Wear glasses to protect eyes.

Acknowledgment

This work was supported by grants from the Toyama Medical Bio-Cluster Project and Hokuriku Innovation Cluster for Health Science of the Ministry of Education, Culture, Sports, Science and Technology, Japan.

References

1. Kohler, G., and Milstein, C. (1975). Continuous cultures of fused cells secreting antibody of predefined specificity. *Nature* 256, 495–497.
2. Kozbor, D., and Roder, J. C. (1981). Requirements for the establishment of high-titered human monoclonal antibodies against tetanus toxoid using the Epstein-Barr virus technique. *J Immunol* 127, 1275–1280.
3. Traggiai, E., Becker, S., Subbarao, K., Kolesnikova, L., Uematsu, Y., Gismondo, M. R., Murphy, B. R., Rappuoli, R., and Lanzavecchia, A. (2004). An efficient method to make human monoclonal antibodies from memory B cells: potent neutralization of SARS coronavirus. *Nat Med* 10, 871–875.
4. McCafferty, J., Griffiths, A. D., Winter, G., and Chiswell, D. J. (1990). Phage antibodies: filamentous phage displaying antibody variable domains. *Nature* 348, 552–554.
5. Winter, G., Griffiths, A. D., Hawkins, R. E., and Hoogenboom, H. R. (1994). Making antibodies by phage display technology. *Annu Rev Immunol* 12, 433–455.
6. Yamamura, S., Kishi, H., Tokimitsu, Y., Kondo, S., Honda, R., Rao, S. R., Omori, M., Tamiya, E., and Muraguchi, A. (2005). Single-cell microarray for analyzing cellular response. *Anal Chem* 77, 8050–8056.
7. Tokimitsu, Y., Kishi, H., Kondo, S., Honda, R., Tajiri, K., Motoki, K., Ozawa, T., Kadowaki, S., Obata, T., Fujiki, S., Tateno, C., Takaishi, H., Chayama, K., Yoshizato, K., Tamiya, E., Sugiyama, T., and Muraguchi, A. (2007). Single lymphocyte analysis with a microwell array chip. *Cytometry A* 71, 1003–1010.
8. Ozawa, T., Kinoshita, K., Kadowaki, S., Tajiri, K., Kondo, S., Honda, R., Ikemoto, M., Piao, L., Morisato, A., Fukurotani, K., Kishi, H., and Muraguchi, A. (2009). MAC-CCD system: a novel lymphocyte microwell-array chip system equipped with CCD scanner to generate human monoclonal antibodies against influenza virus. *Lab Chip* 9, 158–163.
9. Tajiri, K., Kishi, H., Tokimitsu, Y., Kondo, S., Ozawa, T., Kinoshita, K., Jin, A., Kadowaki, S., Sugiyama, T., and Muraguchi, A. (2007). Cell-microarray analysis of antigen-specific B-cells: single cell analysis of antigen receptor expression and specificity. *Cytometry A* 71, 961–967.
10. Jin, A., Ozawa, T., Tajiri, K., Obata, T., Kondo, S., Kinoshita, K., Kadowaki, S., Takahashi, K., Sugiyama, T., Kishi, H., and Muraguchi, A. (2009). A rapid and efficient single-cell manipulation method for screening antigen-specific antibody-secreting cells from human peripheral blood. *Nat Med* 15, 1088–1092.
11. Love, J. C., Ronan, J. L., Grotenbreg, G. M., van der Veen, A. G., and Ploegh, H. L. (2006). A microengraving method for rapid selection of single cells producing antigen-specific antibodies. *Nat Biotechnol* 24, 703–707.
12. Deutsch, M., Deutsch, A., Shirihai, O., Hurevich, I., Afrimzon, E., Shafran, Y., and Zurgil, N. (2006). A novel miniature cell retainer for correlative high-content analysis of individual untethered non-adherent cells. *Lab Chip* 6, 995–1000.
13. Biran, I., and Walt, D. R. (2002). Optical imaging fiber-based single live cell arrays: a high-density cell assay platform. *Anal Chem* 74, 3046–3054.
14. Wrammert, J., Smith, K., Miller, J., Langley, W. A., Kokko, K., Larsen, C., Zheng, N.-Y., Mays, I., Garman, L., Helms, C., James, J., Air, G. M., Capra, J. D., Ahmed, R., and Wilson, P. C. (2008) Rapid cloning of high-affinity human monoclonal antibodies against influenza virus. *Nature 453*, 667–671.

Chapter 12

Analysis of Single Eukaryotic Cells Using Raman Tweezers

Elsa Correia Faria and Peter Gardner

Abstract

Raman Tweezers is a technique that combines optical trapping with Raman spectroscopy and has enabled the spectroscopic analysis of single cells. Applications of this technique include the identification and discrimination of different types of cells, including healthy and non-healthy cells (e.g. cancer cells). In addition, the interaction of cells with stimuli, e.g. drugs, can also be studied on a single-cell basis. Herein, a generic protocol for the analysis of fixed and living single eukaryotic cells is described, including the considerations required to build a Raman Tweezers systems.

Key words: Raman, Optical tweezers, Laser tweezers, Optical trapping Raman microscopy, Laser tweezers Raman microscopy, Laser trapping Raman microscopy, Single-cell analysis, Live-cell analysis, Fixed-cell analysis, Principal component analysis, PCA

1. Introduction

Caution! Before carrying out the work described herein, ensure that you are appropriately trained in the use of lasers, formalin, and cells, and have carried out the appropriate Risk and CoSHH Assessments in accordance with your local health and safety rules. Ensure that you always follow safe operating procedures.

The use of Raman Tweezers (also referred to as optical trapping Raman microscopy, laser tweezers Raman microscopy and laser trapping Raman microscopy in the literature) combines optical trapping (aka laser tweezers) and Raman spectroscopy to enable the manipulation and spectroscopic analysis of single cells, including live cells. Optical trapping allows the cell to be held near the focus of the excitation laser for the duration of the acquisition of the spectrum, for optimal Raman excitation (1).

The Raman spectrum provides information that is molecule specific and can provide information on the biochemical composition

Sara Lindström and Helene Andersson-Svahn (eds.), *Single-Cell Analysis: Methods and Protocols*, Methods in Molecular Biology, vol. 853, DOI 10.1007/978-1-61779-567-1_12, © Springer Science+Business Media, LLC 2012

of the cells. In addition, the spectral bands are sensitive to molecular structure, conformation, and environment, hence allowing investigation of functional groups, bonding types, and molecular conformations. Raman spectroscopy has been increasingly used in medical and biological research, including the investigation of molecular changes associated with disease progression and diagnosis (2–4) and chemical stimuli (5), such as antibiotics. It has also become a well-accepted and widely used method for characterising biological tissues.

Sample preparation and analysis is very simple for Raman Tweezers (3, 4). Cells, live or fixed (e.g. with formalin), can be suspended in aqueous media, such as phosphate-buffered saline (PBS), and individually trapped with the laser and analysed. Analysis in growth medium is usually not recommended, as its components might contribute to the Raman spectrum or might fluoresce, masking the Raman spectrum of the cells. Small amounts of sample can be analysed, without need for staining or modification. The method is mostly non-destructive (depending on wavelength and power of the laser) and non-contaminating. Because of the complex biochemical composition of cells, multivariate chemometric methods, such as Principal Component Analysis (PCA) and Linear Discriminant Analysis (LDA) are usually employed to aid the analysis and interpretation of the data (3, 4).

1.1. Raman Spectroscopy

Any molecule whose vibrations are associated with a change in polarisability of the molecule are Raman active and can give rise to a Raman spectrum. Raman spectroscopy relies on the inelastic scattering of radiation; when electromagnetic radiation falls on a substance, most of it is reflected, absorbed or scattered (Rayleigh scattering). An extremely small fraction of the radiation, however, interacts with the vibrational and rotational states of the molecule. The interacting photons can either impart energy to the molecule or acquire energy from the molecule resulting in radiation being scattered with lower energy (i.e. longer wavelength) or higher energy (shorter wavelength), respectively, than that of the incident radiation. Radiation scattered with lower energy is termed Stokes-shifted Raman and light scattered with higher energy is termed anti-Stokes-shifted Raman.

The relationship between Raman intensity and Raman shift is represented by the Raman spectrum. Only a very small proportion of the incident radiation is Raman scattered (1 in 10^6 to 10^8) and even less is anti-Stokes-shifted Raman, as this requires molecules to be in an excited vibrational state, which normally represents <1% of the molecules at room temperature. Hereunder, reference to Raman scattering is meant as Stokes-shifted Raman. Because of the low Raman scattering efficiency, Raman spectroscopy has only become a useful technique with the invention of lasers. Figure 1 shows the Raman scattering processes.

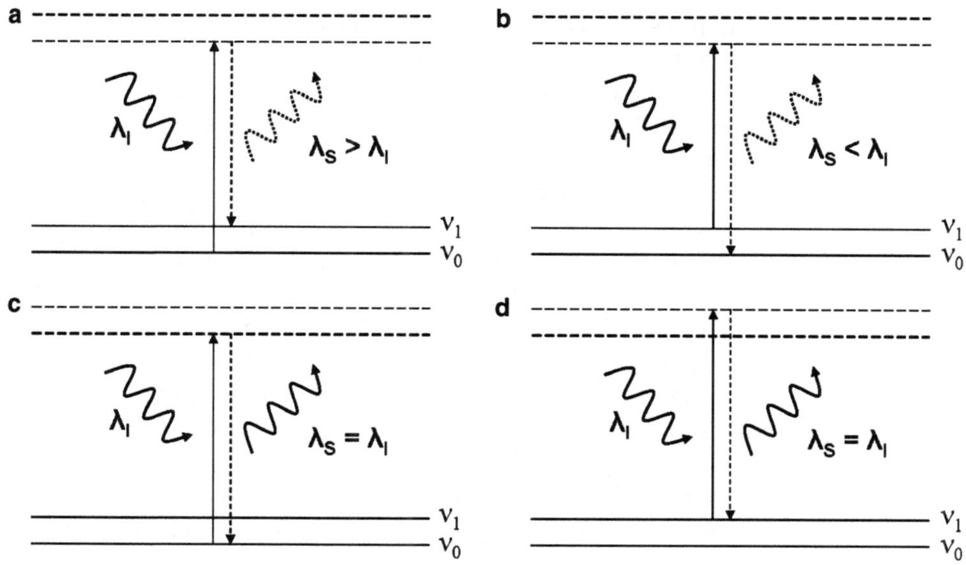

Fig. 1. Representation of the Stokes-shifted (**a**) and Anti-Stokes-shifted (**b**) Raman processes as well as the Rayleigh scattering (**c** and **d**). λ_I is incident (i.e. excitation) wavelength; λ_S is Scattered wavelength; v_0 and v_1 represent the ground vibrational level and the first excited vibrational level, respectively. The horizontal *broken lines* represent virtual energy levels.

Of particular interest are the vibrations termed skeletal vibrations, which involve all the atoms of the molecule, giving rise to a complex spectrum that is typical of the molecule analysed. The energy of these vibrations corresponds to wavenumber shifts in the range 1,400–700 cm^{-1}, which is termed the fingerprint region.

Cells are composed of thousands of molecules of differing types and concentrations, all of which simultaneously contribute to the Raman spectrum of the cell. The resulting spectrum does not allow for identification of individual molecules responsible for a particular band, but the whole spectrum seems to enable the discrimination between different types of cells and with the aid of multivariate techniques give an indication of which type of molecule contribute to the discrimination of the different types of cells.

1.2. Optical Trapping

This technique was pioneered by Arthur Ashkin (6, 7) and enables micron-sized particles to be trapped by a highly focused laser beam. An optical trap is created when a laser beam is brought into tight focus by a high numerical aperture (NA) microscope objective (NA > 1) near a micron-sized particle suspended in a medium of lower refractive index than the particle. The trapping occurs due to the transfer in momentum from the incident beam onto the particle. To achieve optical trapping a microscope (often an inverted microscope) equipped with a high NA objective is used and laser light is guided through the objective onto the sample. With a translation stage, the set-up can be used to move cells in the medium.

1.3. Raman Tweezers

Raman Tweezers combines the above described techniques of Raman spectroscopy and optical trapping (8). Raman Tweezers systems are not yet commercially available; instead they are usually built in-house. Several Raman Tweezers set-ups have been used in the literature (1, 3, 4, 9). Schematics of the different Raman Tweezers configurations are shown in Fig. 2. All set-ups are designed to trap individual cells and excite them with laser light and then measure the Raman scattered light spectrum. The simplest set-up is one which uses an inverted microscope equipped with a high NA objective and a laser that is used simultaneously for trapping the cell and to excite the Raman scattering in a back-scattering mode, as the set-up used by the authors at the Rutherford Appleton Laboratory (Fig. 3) (3). It is this set-up that is described in Subheadings 2 and 3. Other set-ups use separate lasers to trap and excite the Raman spectrum. This is often used if the Raman excitation laser is likely to cause damage to the cells. In such a case, an infrared

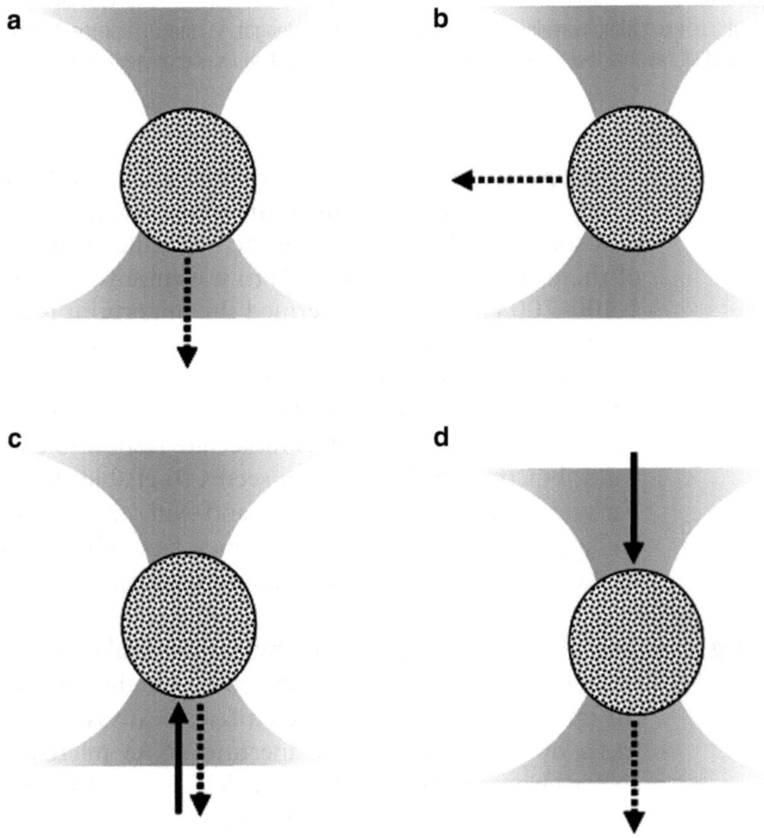

Fig. 2. Possible Raman Tweezers configurations in an inverted microscope. The trapping laser is represented by the *shaded grey area* and the direction of propagation of the measured Raman signal is represented by the *dashed arrows*. The *solid arrows* represent the direction of propagation of the excitation laser light, when excitation is achieved by a laser other than the trapping laser (set-ups **c** and **d**).

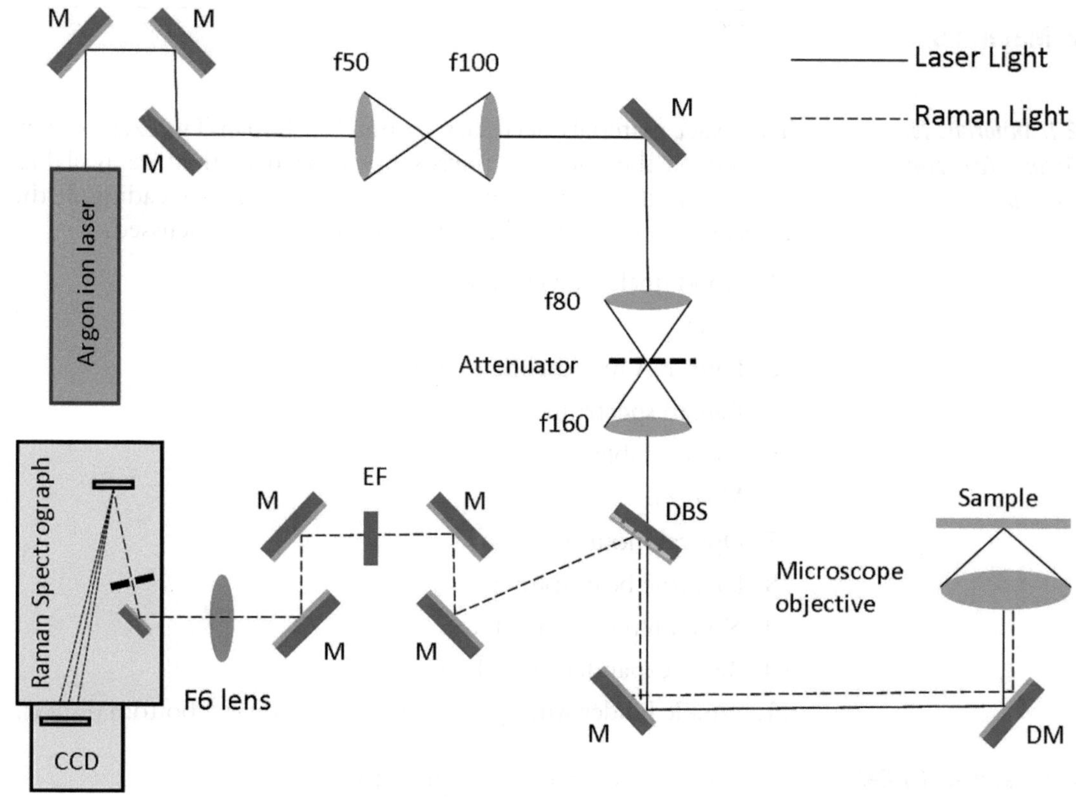

Fig. 3. Raman Tweezers set-up used by the authors at the Rutherford Appleton Laboratory. An Argon-ion laser emitting at 514.5 nm was used both as the trapping and the Raman excitation laser. Raman scattering was detected in back-scattering mode (3). *M* mirror, *DBS* dichroic beam splitter, *EF* edge filter, *DM* dichroic mirror.

laser such as a Nd:YAG or Nd:YVO_4 laser emitting at 1,064 nm can be used as the trapping laser.

The method for analysis of single cells by Raman Tweezers can be summarised in the following steps, which are described in detail in Subheading 3:

1. Build and calibrate Raman Tweezers.
2. Culture/collect/treat cells.
3. If required, fix the cells (acetone, formalin, methanol, SurePath, etc.). See Notes 1 and 2.
4. Optimise measurement parameters, i.e. maximise signal-to-noise ratio (S/N) by determining optimal acquisition time, slit width, and laser power. See Note 3.
5. Trap individual cells and collect their spectra as well as background spectra.
6. Process and analyse data, including background correction, EMSC, and perform multivariate analysis. See Note 4.
7. Interpret data (refer to literature for band assignation), for example ref. 10.

2. Materials

2.1. Materials for Raman Tweezers Set-Up

The exact materials necessary to build a Raman Tweezers system depend on the laser and microscope objectives that are available. Here, only the main components are listed. In Subheading 3, the considerations for the different components are discussed.

1. Good quality inverted microscope.
2. Laser.
3. High NA microscope objective.
4. Raman spectrometer.
5. Optical table.
6. Mirrors.
7. Optical mounts.
8. Dichroic beamsplitter.
9. Notch filter or edge filter.
10. Beam expander set of lenses.
11. Sample holder with quartz bottom (e.g. quartz bottom dishes).

2.2. Materials for Cell Sample Preparation

1. Complete media, pre-warmed to 37°C.
2. 70% ethanol in water (v/v).
3. PBS without Ca^{2+}/Mg^{2+}.
4. 0.25% trypsin/EDTA in HBSS, without Ca^{2+}/Mg^{2+}.
5. Ice.
6. 10% neutral buffered formalin in PBS (or other fixative—see Note 1).

3. Methods

3.1. Building a Raman Tweezers System

Caution! Building a Raman Tweezers System should not be attempted by someone who is not trained and competent in the use of lasers, as serious damage can result from the inappropriate use of lasers, in particular when coupled with focusing optics such as a microscope.

As already mentioned, several configurations are possible. The main components that all configurations have in common, however, are: a high quality microscope equipped with a high NA objective and a translational stage, a Raman spectrometer, a laser for trapping and a laser for Raman excitation (these can be the same laser), means of steering the laser into the objective and onto the samples, means of separating the Raman scattered light from the Rayleigh scattering, means of steering the Raman light onto the spectrometer. A CCD camera that allows visualisation of the field of view via a screen is also useful, so as to avoid laser damage to the eyes resulting from using the eyepiece of the microscope to visualise the sample.

If a commercial Raman microscope is available then it might be possible to transform this into a Raman Tweezer system by using a high NA objective (11). If a Raman microscope is not available, then one can be built by using: a microscope (preferably an inverted microscope), high NA objective, laser, spectrometer, notch filter or edge filter, beam expander and a dichroic beamsplitter. In the single-beam backscattering configuration used by the authors (Fig. 3), the laser light (solid line) has to reach the sample via the objective and the Raman radiation (broken line) is collected via the same path and thus has to be separated from the scattered laser light (Rayleigh) scattering. This is achieved by a high quality dichroic mirror or longpass filter. The system needs to be set up on a vibration free optical table, so that the system is stable enough for the cells to stay in the trap.

3.1.1. The Laser

The laser together with the high NA microscope objective is the key component for trapping and Raman excitation. Continuous wave (CW) lasers that have a Gaussian profile (TEM_{00}) are usually used for Raman Tweezers applications; these include Nd:YAG lasers (532 nm–i.e. frequency doubled 1,064 nm), argon ion lasers (514.5, 488, 457 nm and frequency doubled lines at 257 and 244 nm), and krypton ion lasers (752, 647, 406 nm) lasers. In addition to the 1,064 nm Nd:YAG laser, diode lasers, Ti sapphire, and $Nd:YVO_4$ (914 nm) provide wavelengths in the near infrared region between 690 and 1,000 nm (1). Typical power required for trapping and Raman spectroscopy ranges from a few milliWatt to a few Watt. The laser used for the authors' work was a 500 mW Ar-ion laser emitting at 514.5 nm (3, 4).

When choosing the wavelength of the Raman excitation laser one has to bear in mind that the Raman scattering efficiency is proportional to $1/\lambda^4$. Although lower wavelengths provide greater Raman signal, they also promote fluorescence. Furthermore, one has to take into consideration the potential damage to live cells. 1,064 nm is often used for optical trapping of cells, as at this wavelength photodamage is minimised (if cells are fixed, photodamage does not constitute a problem). At this wavelength, however, the Raman signals are very low and the detectors usually have low response at wavelengths in the infrared. As a compromise, the most commonly used laser is the 785 nm. For efficient trapping and Raman collection, the laser light must slightly overfill the back aperture of the microscope objective; hence, a beam expander set of lenses is required to increase the beam diameter to the right size.

The laser available to the authors at the Rutherford Appleton Laboratory was a 514.5 nm argon-ion laser. Because acquisition times were short (30 s to 1 min), damage to the cells was not significant when analysing live cells. When analysing fixed cells, photodamage was not a consideration.

Caution! Unless the microscope is fitted with appropriate filters that reduce the laser light that reach the eye to a safe level, then the eyepiece of the microscope must not be used for viewing and should be inaccessible. A CCD camera should be used to view the sample instead.

Caution! Before starting any work with lasers you must carry out appropriate Risk Assessment and be appropriately trained. You must follow all safety precautions in accordance with your local health and safety rules and ensure that you carry out your work in a safe manner to avoid damage to your eyes. The risk of damage to the eyes is further increased by coupling a laser to a microscope. Appropriate Safe Operating Procedures must be in place before work can commence.

3.1.2. The Objective

The objective is together with the laser fundamental for trapping. The key requirement for stable trapping at the focal point is that the objective must have a high NA (typical NA is between 1.2 and 1.4). Objectives of such high NA are immersion objectives, with both oil and water immersion types being used. Water immersion objectives usually allow trapping of particles further into the medium than oil immersion objectives. The objective used by the authors was a Leica, NA = 1.2 water immersion objective.

If two separate lasers are to be used for trapping and exciting Raman scattering, then it is important to select an objective that corrects for chromatic aberrations, to ensure that the foci of the two lasers are as close as possible.

3.1.3. The Dichroic Beamsplitter, Edge Filter, and Notch Filter

As can be seen in Fig. 3, the laser light and the Raman scattetring share the optical path. A dichroic beamsplitter allows the laser light to pass and reflects the backscattered Raman light. In addition to the Raman light the backscattered light still contains some Rayleigh scattering which needs to be removed before the light reaches the spectrometer. This is achieved with the use of either a notch filter or a longpass filter (i.e. an edge filter that transmits light of longer wavelength than that of the excitation laser). A notch filter blocks the laser wavelength, while the longpass filter blocks the wavelengths below the laser wavelength.

3.1.4. The Spectrometer

The majority of spectrometers used for Raman spectroscopy are grating spectrometers, in which the Raman scattered light is dispersed into its constituent wavelengths by a grating in the spectrometer and these are detected by a CCD array. Gratings are characterised by their blaze wavelength (i.e. wavelength at which it is most efficient), which defines the wavelength range for which they can be used. The grating used in the measurements will depend on the laser used for the Raman excitation.

Commonly, shifts between 700 cm^{-1} and 1,800 cm^{-1} are measured. The Raman wavelength range that this corresponds to can be determined by the following:

$$\lambda_R = \frac{\lambda_I}{1 - \lambda_I \times \overline{v}_S \times 10^{-7}}$$

where: λ_R is the wavelength of the Raman radiation in nm; λ_I is the wavelength of the excitation light in nm; $\overline{v}_S$ is the shift wavenumber in cm^{-1}.

Hence, the maximum and minimum wavelengths that need to be detected are determined by: $\lambda_R = \frac{\lambda_I}{1 - \lambda_I \times 1800 \times 10^{-7}}$ and $\lambda_R = \frac{\lambda_I}{1 - \lambda_I \times 700 \times 10^{-7}}$, respectively.

The spectral resolution achieved (i.e. the minimum shift that can be resolved) is dependent on the entrance slit of the spectrometer and the groove density of the diffraction grating (1, 12). Increasing the groove density and/or reducing the entrance slit will result in increased spectral resolution. However, increasing the groove density will also reduce the spectral range that can be detected, and reducing the slit width reduces the light that reaches the spectrometer, thus reducing the signal-to-noise ratio (S/N).

3.1.5. Other Components

In addition to the components described above, high quality mirrors and lenses are required to guide and focus the laser beam to the sample through the microscope and the Raman radiation from the sample to the spectrometer. The sample should be placed in a sample holder made of a material that does not fluoresce, such as quartz, and be of suitable thickness for the objective (this should be indicated on the objective).

3.1.6. Calibration

Trapping and acquisition of Raman spectra should be first attempted using microbeads (e.g. 10 μm in diameter). Polystyrene beads are quite useful, as polystyrene yields a very characteristic spectrum. In addition, the polystyrene spectrum can be used to calibrate the wavenumber shift.

3.2. Analysis of Single Cells Using Raman Tweezers

3.2.1. Preparation of Cells for Analysis

Caution! Before working with cells ensure that you have carried out the appropriate risk and CoSHH assessment prior to commencement of the work. Follow your local health and safety rules when carrying out the work.

If cells are non-adherent, go straight to the relevant Analysis section. (For further detail on cell culture technique refer to a cell culture manual such as the ECACC Handbook (13)).

1. Grow adherent cells in T25 flasks until ca. 80% confluent.
2. Check health and degree of confluency of the cells under a microscope.
3. Pre-warm media and PBS to 37°C.
4. Following good sterile cell culture practice. Remove spent medium.
5. Wash the cell monolayer with 5 mL PBS without Ca^{2+}/Mg^{2+}. Repeat if cells are strongly adherent.
6. Add 1 mL of trypsin and tilt and rotate flask to ensure that the whole monolayer is covered in trypsin. Decant the excess trypsin.
7. Place flask in incubator for 2–10 min checking regularly under a microscope to determine when all cells are detached and floating. If some cells are still attached the sides of the flask can be tapped gently to aid detachment.
8. Re-suspend the cells in 6 mL of fresh complete medium (see Note 5).
9. Transfer the required number of cells to a new labelled flask containing the necessary volume of pre-warmed medium to achieve required cell seeding density. (You might want to prepare two flasks at this stage just for redundancy).
10. Place cells in incubator and incubate in the appropriate conditions for the cell line.
11. Keep the remainder of the cells for analysis.

3.2.2. Analysis of Non-fixed Cells

1. Divide the cells suspended in medium into 1 mL aliquots (one sixth of the cells in the T25 flask) and keep in falcon tubes on ice until required for analysis (no longer than 8 h).
2. Just before analysis, centrifuge one 1 mL aliquot at 1,500 rpm (150 x g) for 5 min.
3. Decant supernatant leaving the pellet of cells in the bottom of the falcon tube.
4. Add 1 mL of PBS without Ca^{2+}/Mg^{2+} and resuspend cells.
5. Place 200–300 μL of cells in the sample holder.
6. Place remaining cells on ice (cells should not be kept in PBS for longer than 2 h).
7. Place immersion fluid (oil or water depending on the objective used) and place sample holder with sample on the microscope stage and adjust the focus.
8. Move the stage laterally until a cell is found and is positioned in the centre of the trap.
9. Adjust the focus onto the cell.
10. Trap the cell and if the cell is too close to the surface of the sample holder lift it slightly above.

11. Measure a spectrum.
12. Optimise the acquisition time and the laser power to optimise the signal-to-noise ratio (i.e. to obtain the greatest signal-to-noise ratio. This can be estimated using the amide I band around 1,650 cm^{-1}). (See Note 3) (The authors have used 1 min acquisition time with a laser power of 400 mW of a 514.5 nm laser, 100 mW delivered to the cell as measured by a power meter placed at the sample position on the microscope stage).
13. Once the optimum acquisition time and laser power has been determined keep the parameters constant for the duration of the analysis of all the samples.
14. Find and trap a new cell and acquire a spectrum.
15. Maintaining the same conditions, release the cell, and acquire a background spectrum, i.e. a spectrum of the solution without a cell at exactly the same height. If the trapping depth (i.e. the depth at which the cells are analysed) does not change, then the background only needs to be acquired once in every five spectra. If the trapping depth does change, then a spectrum has to be acquired after each cell spectrum that is acquired at a different depth.
16. Repeat steps 14 and 15 to obtain spectra of other cells. A minimum of 20 cells should be analysed for the analysis to be statistically significant.

3.2.3. Analysis of Fixed Cells

If the analysis of a particular sample of cells is envisaged to take longer than 1–2 h or if it is necessary to keep the cells for further analysis, then the cells should be fixed. Several fixation protocols exist, including formalin fixation (see Note 1), methanol fixation, cytological fixatives, such as SurePath, etc. The aim of fixation is to preserve the biochemical and structural integrity of cells, while minimising any possible biochemical or structural changes (14).

Formalin fixation is probably the most widely used since, although the spectra are slightly altered compared to the live cell spectra these changes are well documented and are reasonably consistent from cell to cell (see Note 1). In addition, fixation allows cells to be stored at room temperature. Formalin fixation is the method that is described herein.

Caution! Formalin is toxic. Ensure that you have carried out the appropriate risk and CoSHH assessment prior to commencement of the work. Follow your local health and safety rules.

1. To fix cells, start from step 8 of Subheading 3.2.1.
2. Centrifuge the cells suspended in complete media at 1,500 rpm (150 x g) for 5 min and decant the supernatant.
3. Resuspend cells in 5 mL of PBS and centrifuge at 1,500 rpm (150 x g) for 5 min; decant the supernatant.

4. Resuspend cells in 1 mL of 10% formalin (i.e. 4% formaldehyde, though often erroneously referred to as 4% formalin, see Note 2). These cells can be stored at room temperature until required for analysis and for extended periods.
5. Analyse cells suspended in formalin as for PBS suspended cells. Follow steps 5–16 of Subheading 3.2.2.

3.3. Data Analysis

The authors have carried out principal component analysis (PCA) on their data using MATLAB (3, 4). Other software such as SPSS can also be used, but is not discussed in this chapter.

1. To analyse the spectral data in MATLAB, the data files have to be in a format that the software is able to read, such as comma separated values (.csv), tab-delimited (.txt) or .spc. If the spectral data are not in a format readable by MATLAB, then they will need to be converted into such a format.
2. Subtract the appropriate background (i.e. the spectrum obtained without a cell in the trap) from each cell spectrum.
3. Select the fingerprint region 700–1,750 cm^{-1} for further processing.
4. Perform extended multiplicative scattering correction (EMSC) using for each cell type the average spectrum for that cell type as the reference spectrum.
5. Carry out vector normalisation and mean centering.
6. Perform quality control of the spectra and remove spectra that have a low S/N and a low signal intensity.
7. Carry out PCA analysis on the remaining spectra.
8. Plot the scores plots in 2D and/or 3D.
9. Plot the loading plots for the principal components of interest.
10. For tentative band assignation refer to the literature, e.g. (10).

4. Notes

Fixation: If fixation of the cells is not desired, then the cells can be suspended and analysed in PBS. PBS is used because it is suitable for use with cells, contributes little to the Raman spectrum and is of lower refractive index then the cells and hence is a suitable medium for trapping cells. However, it must be noted that when cells are removed from their culture medium and left unfixed, autolysis and putrefaction will quickly set in. Cells should therefore not be kept in PBS for longer than 1–2 h.

Autolysis is a complex process of self-destruction caused by inter-cellular enzymatic processes, which include denaturation of proteins dephosphorylation of phospholipids, nuclear fragmentation, chromatin compaction and cytoplasmic condensation (3, 15), which eventually lead to liquefaction of the cells. Putrefaction on the other hand is caused by bacterial attack. Fixation of the cells will stabilise the proteins in the cells and thus delay the processes of autolysis and putrefaction. By fixing cells, these can be preserved in as close as life-like state as possible.

There are many fixation methods available that are used in hystology, formalin being one of the most commonly used. This has also become the most commonly used fixative for spectroscopic analysis. Although the spectra of fixed cells are slightly altered compared to the spectra of live cell, these changes are well documented and are reasonably consistent from cell to cell (16–21). In addition to the lower effects, it requires a short fixation time and it can be used for long-term storage without causing deleterious effects on cell morphology and adequately preserving nuclear and cytoplasmic detail. Hence, formalin fixation is the method described herein.

Formalin acts as fixative by cross-linking proteins, i.e. by reacting with primary and secondary amine groups of proteins forming chemically stable methylene bridges (14), which are insoluble rigid lattices that prevent the escape of cellular constituents. Lipids are also preserved by the addition of methylene glycol across the double bonds of lipids, but formalin has no effect on carbohydrates. Formalin can also cross-link DNA by the formation of protein-DNA adducts (22).

Formalin Nomenclature: There is some confusion in the literature about the nomenclature and interchangeable use of formalin and formaldehyde. Formalin is the saturated aqueous solution of formaldehyde (which exists in solution in the hydrated form termed formalin or methylene glycol). This saturated solution is 40% (w/v) in formaldehyde and can be prepared from paraformaldehyde, which is a polymer of formaldehyde. The solution used for fixation is often erroneously referred to as 4% formalin; it is actually 10% formalin (i.e. 10% solution of the 40% saturated formaldehyde solution) or 4% formaldehyde.

Signal-to-Noise: Noise can be defined as the "*intensity in the spectrum from which useful information is not being derived*" (12).

In single-cell Raman spectroscopy where differences in the spectra of different cells are very small, it is essential to maximise the signal-to-noise ratio (S/N), which is defined as the ratio of the true signal intensity to the root-mean-square (RMS) fluctuation of the signal from its true value (noise). Spectral noise can be subdivided into two types; additive (background) and multiplicative noise. Additive noise is always present, i.e. in the presence and the absence of a sample, and comprises dark current as well as contributions from the blank (when no sample is present, i.e. the suspension medium). Additive noise is independent of the signal strength, while multiplicative noise increases with signal strength (1). Additive noise can be eliminated from the sample spectrum by subtracting the blank spectrum from the sample spectrum.

The most dominant form of noise in Raman spectroscopy is shot noise (12), which is due to the quantum nature of matter and photons, and whose magnitude is inversely proportional to signal strength (23). Thus the S/N can be increased by increasing the spectrum acquisition time. Johnson noise results from the thermal motion of electrons in resistive elements in electric circuits, and is independent of signal magnitude (23). Flicker noise, which is largely associated with the laser source and is dependent on the signal magnitude (12, 24). Therefore, if flicker is the dominant noise contribution, increasing integration time (i.e. acquisition time) will not increase the S/N; to minimise the contribution of flicker noise the laser has to be stable. Once the set-up has been built, the variables that can be optimised to maximise S/N are: power, acquisition time and slit width. These must be optimised before analysis begins, and once found must be kept constant for the duration of the experiment.

Data Pre-processing: Before analysis and interpretation of data can take place it is necessary to quality control the spectra to remove spectra with poor S/N and spectra with contributions from cosmic rays (although the cosmic ray contribution can be eliminated mathematically). This is then followed by pre-processing of the data to remove: sample-to-sample intensity variations such as differences in cell size, variation in laser power, trapping depth, etc.; undesired spectral artefacts such as scattering and fluorescence; smooth out noise; convert the original data scale into a more convenient scale. The pre-processing methods used by the authors for analysis of data obtained using Raman Tweezers are as follows: Extended Multiplicative Signal Correction (EMSC), vector normalisation, and mean centring followed by PCA.

Extended Multiplicative Signal Correction: The EMSC method was first developed by Martens and Stark to correct for additive (baseline) and multiplicative effects in near-infrared spectroscopy (25), but has since found applications in correcting Raman spectra of biological samples (26–29). EMSC builds on the multiplicative signal correction (MSC) method, in which the additive and multiplicative effects are first estimated by least squares regression with respect to a reference spectrum. The corrected spectrum is then obtained by subtracting every original value by the additive effect and dividing by the multiplicative effect. In addition, EMSC also corrects for wavelength dependent spectral variations from sample to sample. The effect of EMSC on Raman spectra is illustrated in Fig. 4. Differences in the baselines of spectra shown in A have been removed in B.

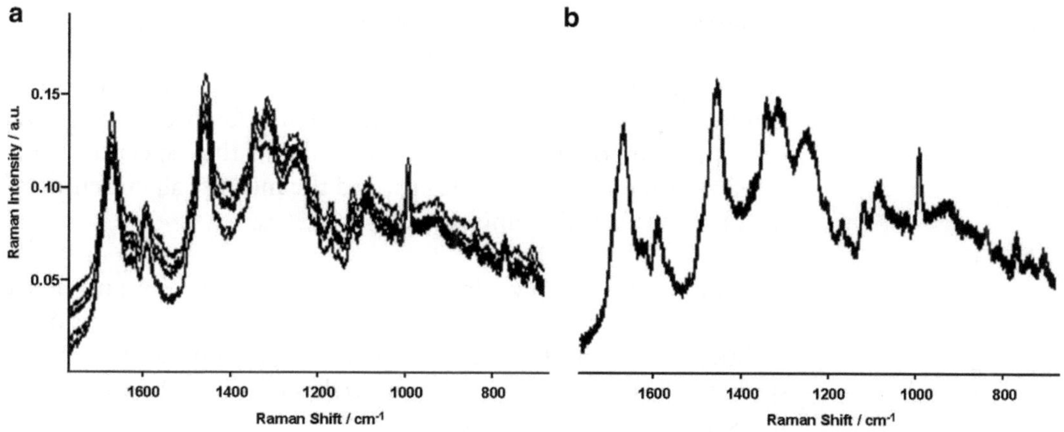

Fig. 4. Uncorrected Raman spectra (**a**) and EMSC corrected spectra (**b**) (35).

Vector Normalisation: Vector normalisation is used to remove multiplicative effects, such as variations in spectral intensities due to cell size, variation in laser power, trapping depth, etc. (30–32). Normalisation is performed by dividing the intensity in a spectrum at each wavenumber by the square root of the sum of the squares of the intensity at each wavenumber of that spectrum. This leads to the sum of the square of the normalised intensities being equal to one (27). Vector normalisation is normally preceded by mean centring.

Smoothing: The purpose of smoothing is to reduce the amount of noise in a spectrum, and thus increase the signal-to-noise ratio (S/N). Smoothing always results in a reduction of useful information and resolution. Hence, smoothing should only be used with caution and if S/N cannot be improved in the raw spectra by optimising measurement parameters, such as laser power, trapping depth, acquisition time and slit width.

There are a number of different methods of smoothing available, but one of the most commonly used is the Savitzky–Golay (S–G) algorithm (33). In this method, each point of the smoothed spectrum is a weighted average of the point and an equal number of adjacent points on each side of the point in the original spectrum so that the point in the smoothed spectrum is weighted average of an odd number or points (these can be specified by the user) (34). The weight of each point is fitted by a polynomial curve, which can be a quadratic, cubic, or quartic and with the number of points can be specified by the user. The degree of smoothing depends on the number of points used in the smoothing process; the higher the number of points, the greater the degree of smoothing and the lower the spectral resolution.

Mean Centring: Mean centring involves subtracting the mean spectrum of the spectral data set from each spectrum, i.e. for each wavelength of a spectrum subtract the mean intensity at that wavelength of all spectra. This is a required step before carrying out principal component analysis. This results in the "spectra" being spread symmetrically around zero, and the mean of all "intensities" for each wavelength being zero.

Neutralisation of Trypsin: Serum neutralises trypsin and hence the monolayer of cells needs to be thoroughly washed prior to addition of trypsin. Prolonged exposure to trypsin can damage the surface cell receptors and therefore cells should not be exposed to trypsin longer than necessary for cells to detach (no longer than 10 min). When incubating cell monolayers with trypsin, regularly check for progress in the detachment process. Once cells are detached, trypsin must be neutralised with serum (this can be by using the medium containing serum) before seeding the cells in new flasks; otherwise, cells will not attach.

References

1. Snook RD, Harvey TJ, Correia Faria E, and Gardner P. (2009) Raman tweezers and their application to the study of singly trapped eukaryotic cells. *Integrative Biology 1*, 43–52.
2. Rusciano G, De Luca AC, Pesce G, and Sasso A. (2008) Raman tweezers as a diagnostic tool of hemoglobin-related blood disorders. *Sensors 8*, 7818–7832.
3. Harvey TJ, Faria EC, Henderson A, Gazi E, Ward AD, Clarke NW, Brown MD, Snook RD, and Gardner P. (2008) Spectral discrimination of live prostate and bladder cancer cell lines using Raman optical tweezers. *Journal of Biomedical Optics 13.*
4. Harvey TJ, Hughes C, Ward AD, Faria EC, Henderson A, Clarke NW, Brown MD, Snook RD, and Gardner P. (2009) Classification of fixed urological cells using Raman tweezers. *Journal of Biophotonics 2*, 47–69.
5. Moritz TJ, Polage CR, Taylor DS, Krol DM, Lane SM, and Chan JW. (2010) Evaluation of Escherichia coli cell response to antibiotic treatment by use of Raman spectroscopy with laser tweezers. *Journal of Clinical Microbiology 48*, 4287–4290.
6. Ashkin A. (1970) Acceleration and Trapping of Particles by Radiation Pressure. *Physical Review Letters 24*, 156–159.
7. Ramser K, and Hanstorp D. (2010) Optical manipulation for single-cell studies. *Journal of Biophotonics 3*, 187–206.
8. Thurn R, and Kiefer W. (1984) Raman-Microsampling Technique Applying Optical Levitation by Radiation Pressure. *Applied Spectroscopy 38*, 78–83.
9. Creely CM, Singh GP, and Petrov D. (2005) Dual wavelength optical tweezers for confocal Raman spectroscopy. *Optics Communications 245*, 465–470.
10. Movasaghi Z, Rehman S, and Rehman IU. (2007) Raman spectroscopy of biological tissues. *Applied Spectroscopy Reviews 42*, 493–541.
11. Neugebauer U, Bocklitz T, Clement JH, Krafft C, and Popp J. (2010) Towards detection and identification of circulating tumour cells using Raman spectroscopy. *Analyst 135*, 3178–3182.
12. Pelletier MJ. (1999) in *Analytical Applications of Raman Spectroscopy* (J., P. M., Ed.) pp 23–25, Blackwell Science, Michigan.
13. ECACC (2010) *Fundamental Techniques in Cell Culture: Laboratory Handbook*, 2nd ed., Sigma-Aldrich.
14. Kiernan JA (1999) *Histological and Histochemical Methods: Chapter 2* Butterworth Heinemann, Oxford.
15. Leong A. (2000) Fixation and fixatives in *Laboratory Histopathology: A complete Reference* (Leong, A., Ed.) pp 12-43, Woods and Ellis, London.
16. Draux F, Gobinet C, Sulé-Suso J, Trussardi A, Manfait M, Jeannesson P, and Sockalingum

GD. (2010) Raman spectral imaging of single cancer cells: Probing the impact of sample fixation methods. *Analytical and Bioanalytical Chemistry 397*, 2727–2737.

17. Meade AD, Clarke C, Draux F, Sockalingum GD, Manfait M, Lyng FM, and Byrne HJ. (2010) Studies of chemical fixation effects in human cell lines using Raman microspectroscopy. *Analytical and Bioanalytical Chemistry 396*, 1781–1791.
18. Mariani MM, Lampen P, Popp J, Wood BR, and Deckert V. (2009) Impact of fixation on in vitro cell culture lines monitored with Raman spectroscopy. *Analyst 134*, 1154–1161.
19. Chan JW, Taylor DS, and Thompson DL. (2009) The effect of cell fixation on the discrimination of normal and leukemia cells with laser tweezers Raman spectroscopy. *Biopolymers 91*, 132–139.
20. Hastings G, Wang R, Krug P, Katz D, and Hilliard J. (2008) Infrared microscopy for the study of biological cell monolayers. I. Spectral effects of acetone and formalin fixation. *Biopolymers 89*, 921–930.
21. Ó Faoláin E, Hunter MB, Byrne JM, Kelehan P, McNamara M, Byrne HJ, and Lyng FM. (2005) A study examining the effects of tissue processing on human tissue sections using vibrational spectroscopy. *Vibrational Spectroscopy 38*, 121–127.
22. Hayat MA (2000) *Principles and Techniques of Electron Microscopy* Cambridge University Press, Cambridge
23. Skoog DA, and West DM (1981) *Principles of Instrumental Analysis: Chapter 3*, Holt-Saunders, Tokyo.
24. Stevenson CL, and Vo-Dinh T. (1996) Signal expressions in Raman spectroscopy in *Modern techniques in Raman spectroscopy* (Laserna, J. J., Ed.) pp 1-39, John Wiley & Sons, Chichester.
25. Martens H, and Stark E. (1991) Extended multiplicative signal correction and spectral interference subtraction: New preprocessing methods for near infrared spectroscopy. *Journal of Pharmaceutical and Biomedical Analysis 9*, 625–635.
26. Heraud P, Wood BR, Beardall J, and McNaughton D. (2006) Effects of pre-processing of Raman spectra on in vivo classification of nutrient status of microalgal cells. *Journal of Chemometrics 20*, 193–197.
27. Afseth NK, Segtnan VH, and Wold JP. (2006) Raman spectra of biological samples: A study of preprocessing methods. *Applied Spectroscopy 60*, 1358–1367.
28. De Gelder J, De Gussem K, Vandenabeele P, De Vos P, and Moens L. (2007) Methods for extracting biochemical information from bacterial Raman spectra: An explorative study on Cupriavidus metallidurans. *Analytica Chimica Acta 585*, 234–240.
29. De Gussem K, Vandenabeele P, Verbeken A, and Moens L. (2007) Chemotaxonomical identification of spores of macrofungi: Possibilities of Raman spectroscopy. *Analytical and Bioanalytical Chemistry 387*, 2823–2832.
30. Lasch P, Diem M, and Naumann D. (2004) FT-IR microspectroscopic imaging of prostate tissue sections. *Proc. of SPIE 5321*, 1–9.
31. Sahu RK, Argov S, Salman A, Zelig U, Huleihel M, Grossman N, Gopas J, Kapelushnik J, and Mordechai S. (2005) Can Fourier transform infrared spectroscopy at higher wavenumbers (mid IR) shed light on biomarkers for carcinogenesis in tissues? *Journal of Biomedical Optics 10*, 054017-1-054017–10.
32. Salman A, Ramesh J, Erukhimovitch V, Talyshinsky M, Mordechai S, and Huleihel M. (2003) FTIR microspectroscopy of malignant fibroblasts transformed by mouse sarcoma virus. *Journal of Biochemical and Biophysical Methods 55*, 141–153.
33. Savitzky JA, and Golay MJE. (1964) Smoothing and Differentiation of Data by Simplified Least Squares Procedures *Anal. Chem. 36*, 1627–1639.
34. Ferraro JR, Nakamoto K, and Brown CW. (2002) Chapter 5 – Analytical Chemistry, in *Introductory Raman Spectroscopy* pp 267–293, Academic Press, San Diego.
35. Harvey TJ. (2008) Development of vibrational spectroscopic cytology for prostate cancer diagnosis, *School of Chemical Engineering and Analytical Science*, PhD Thesis The University of Manchester, Manchester.

Chapter 13

Single-Cell Microinjection Technologies

Yan Zhang

Abstract

Single-cell microinjection, as a mechanical delivery tool, has been used for transferring substances into transfection or infection challenging cells. Here, we discuss the advantages and applications of microinjection, list the materials needed for performing microinjection experiments, and describe the methods of single-cell microinjection into suspended and attached cells.

Key words: Microinjection, Attached cells, Suspended cells, Single-cell analysis

1. Introduction

As a mechanical delivery method, single-cell microinjection is widely applied to transduce proteins, peptides, cDNAs, interference RNAs, sperms, and large molecule nondiffusible drugs (1, 2). Compared with chemical and viral delivery methods, microinjection has relatively low cytotoxicity. Microinjection is especially useful in transferring materials into cells resistant to chemical transfection, such as primary human, rat, and mouse neurons (1–8). Theoretically, the transduction efficiency of the cells successfully injected and survived injection is 100%. During microinjection, the effects induced by the injected substances could be largely isolated given that, theoretically, the injected substances are the only independent variable in well-controlled experiments. More than one substance can be injected into different groups of cells in one culture dish, where the uninjected cells serve as built-in controls, making the results more compatible. In the case of delivery proteins and peptides, microinjection requires much less protein preparation. This is especially important for the experiments with less abundant proteins, expensive recombinant proteins, and synthesized peptides (see Note 1). Microinjection also delivers with the precise controls

Sara Lindström and Helene Andersson-Svahn (eds.), *Single-Cell Analysis: Methods and Protocols*,
Methods in Molecular Biology, vol. 853, DOI 10.1007/978-1-61779-567-1_13, © Springer Science+Business Media, LLC 2012

of volume and timing. The location of delivery into either the cytosol or the nucleus is also controlled by microinjection, which is difficult to achieve in many cases with electroporation, carrier peptide delivery, liposome-mediated transfection, or virus-based infection. The injected cells can be identified by the coinjection of a cell membrane impermeable marker dye such as dextran Texas red (DTR), Alexa488 or by the protein conjugated to fluorescence, which ensures to recognize the injected cells. As a single-cell technique, immunofluorescence, in situ cell death detection, such as terminal deoxynucleotidyl transferase-biotin dUTP nick-end labeling (TUNEL), subcellular location or translocation and single cell polymerase chain reaction results are the common follow-up assays for microinjection experiments (4–9).

Microinjection has been applied to study human embryo (10) and mitotic cells (see Note 2) (11). Microinjection is also widely used in transferring RNA interference (12–16), deliver RNAs (17, 18), neutralizing antibodies (14, 19–21), dominant negative mutants (22, 23), sperms (24, 25), and nanoparticles (26) into cells. Also microinjection is useful in transgenic animal production (27, 28) and in vitro fertilization in medical clinics (10). Recently, single-cell microinjection has been used to study gap junction communication (29) and viral transcriptional regulatory protein domains (30). In particular, microinjected cells can be recognized by a marker dye and monitored with electrophysiological approach (31, 32). Microinjection has been extensively used in our studies with primary cultured human or rodent neurons that are difficult to transfect or infect. Proteins (e.g., recombinant caspases, Bax, Hsp70, neutralizing antibodies), peptides (e.g., amyloid β), and various cDNA constructs are delivered into human or rat neuronal cytosol (4–8, 32–38). In addition, cDNA constructs are also injected into the nuclei of a free moving protozoan euplotes (unpublished results).

2. Materials

1. *Microscope, micromanipulator, and microinjector.* An inverted microscope with adjustable magnification of at least 200× is required, since the noninverted microscope greatly limits the working distance between the lens and objectives (3). For nuclear injection, a magnification of 400× is necessary. The pseudo three-dimensional image provided by the Nomarski or Hoffman optical systems may be helpful in recognizing the nuclei during nuclear injection (27). Generally, the illumination from halogen bulbs of microscopes is good enough for the cytosolic injections (10). If nuclear injection is desired, a fiber optic light source is necessary to visualize the boundary of

the nuclei (10, 27). Two micromanipulators that allow fine and smooth three-dimensional movements manually or automatically are necessary to precisely position the holding and injection needles. The micromanipulators that reduce most of the vibrations from hands, body and environment is preferred (10). A microinjector controlling the compensation pressure, injection pressure, and injection time, such as FemtoJet (Eppendorf, Hamburg, Germany), is required.

2. *Injection needles, glass capillaries and puller.* In our experiments, the injection needles are pulled from the thin-walled MTW100F-4 borosilicate glass capillaries with microfilament (World Precision Instruments, Sarasota, FL, OD = 1.0 mm, ID = 0.5 mm). Needles are pulled by a micropipettor puller. The puller electrically heats up the glass capillary with a heating element. Then a horizontal linear force pulls the heated glass apart to produce two tapered needles (3). During the end stage of pulling, a gas jet blows nitrogen or compressed air to break the glass capillary. Therefore, pulling force, pulling velocity, heating temperature and gas blowing timing are the critical parameters for the shape and tip diameter of injection needles. The needles used for cytosolic microinjection in our study ideally are gradually tapered with the distance between the shaft and the tip as 0.4–0.6 cm and the approximate tip diameter as 0.5 μm. Commercially available ready-to-use pipettes such as the Femtotips (Eppendorf, Hamburg, Germany) are also available.

3. *Marker dye.* DTR (3,000 MW; Molecular Probes, Carlsbad, CA) or Alexa488 (Molecular Probes, Carlsbad, CA) is injected as a fluorescent marker to recognize the injected cells. DTR or Alexa488 is dissolved at 100 mg/ml in phosphate buffered saline (PBS: 140 mM NaCl, 2.7 mM KCl, 4.3 mM Na_2HPO_4, 1.5 mM KH_2PO_4, pH 7.4) and stored at −20°C. Mix the marker dye with the materials to be injected immediately before the injection. Avoid frequently freezing and thawing stock solution.

3. Methods

3.1. Establishing the Parameters for Needle Pulling

To establish the parameters of the puller, the proper glass melting temperature is first determined. In general, higher glass temperature, typically ranges from 300 to 700°C, results in greater plasticity and finer-tipped needles with a longer tapered area. Lower melting temperature is desired for pulling holding needles. The pulling force and velocity are also important for needle shape. In general, greater pulling force and velocity result in finer tips.

3.2. Preliminary Control Experiments

1. *Determine the injection parameters.* Set the injection pressure in the range of 50–100 hPa. Set the compensation pressure at approximately 50% of the injection pressure. The compensation pressure is required to balance the capillary forces caused by the fine tip of the injection needle. Due to capillary forces, liquid from the culture dish tends to flow up into the injection needle and dilute the injection solution inside of the needle. To prevent this, the compensation pressure is adjusted to push the solution inside of the needle to the tip area and ready to be injected. A value which ensures a continuous trace amount of discharge out of the needle tip should be determined before experiments by using a visible dye such as fast green. Set the injection time as 0.1–0.5 s.
2. *Determine the injected volume.* The injection pressure, compensation pressure, and injection time determine the volume of each injection. Fill one needle with the injection solution using a Microloader loading tip (Eppendorf, Hamburg, Germany) and avoid air bubbles inside of the needle. The injection volume can be determined by loading the needle with 0.1–0.2 μl of fast green and injecting the solution into medium. Count the number of ejections required to finish up the solution. The injection volume can be calculated by dividing the total volume loaded in the needle by the number of injections. Based on the injection volume and solution concentration, the actual amount of material delivered for each shot of injection can be calculated. Cytosolic volumes vary from different cell types, and this should be taken into account when calculating the dosage of injection in each cell.
3. *Determine the holding pressure.* If free moving or suspended cells are about to be injected, a holding needle is required to limit the movement of cells. The negative pressure can be generated by a manual microinjector such as AirTram (Eppendorf, Hamburg, Germany). Adjust the handle to produce a negative pressure to suck the cell to the tip of the holding needle. Be careful not to induce too much negative pressure to damage the held cell.
4. *Optimize the concentration of marker dye.* The concentration of marker dye, such as DTR or Alexa488, used for microinjection needs to be determined by injecting various doses of dye in PBS, ranging from 10 to 1,000 μg/ml, into cells, and counting the number of dye positive cells after injection.
5. *Determine the toxicity of the injected buffers.* All materials and solutions for injections should be sterilized. The injected buffer may contain a number of chemicals that could be toxic to cells. Solutions containing detergents such as NP40, Brij3 or 5, Triton X-100, glycerol above 20%, bacteriostatics such as sodium azide and salt solutions above 400 mM are not recommended

for microinjection. Cells are unable to balance intracellular pH against strong buffers. The cells should be microinjected with a buffer which induces the least perturbation. Therefore, always check the possible toxicity of the injection solution components. The toxicity of buffer is tested by injecting the buffer containing the predetermined concentration of marker dye into cells and observing cell viability after incubation.

6. *Determine cellular viability after the injection.* Measure cell viability using TUNEL staining after various time of injection ranging from 0 to 20 days to confirm the injection itself does not induces significant cell death (see Note 3).

3.3. Microinjection

1. Prepare the injection solutions by mixing the marker dye and the injected materials with buffers at the desired concentration.
2. Place the injection needle in the micromanipulator holder. Place the holding needle in the micromanipulator holder at the other side if injecting suspended or free moving cells.
3. Carefully take out cells with culture medium. Place the Petri dish onto the stage of the microscope. Adjust focusing until the structure of cytosolic and nuclear areas appears.
4. Select an area containing morphologically healthy cells. If the cells are attached, it is preferable to inject cells in one area, which facilitates the detection of the injected cells.
5. Apply negative pressure to suck one cell at the tip of the holding needle. Position the injection needle on the top of the cell using the manual micromanipulator and joystick. For cytosolic injection of cells, place the injection needle just on the top of the cell surface.
6. Carefully lower down the injection needle by slowly turning the vertical control of the joystick to let the tip penetrate into the cell membrane. This step is critical for a successful injection and cell survival from the injection.
7. Press the "inject" button of the microinjector to finish the injection (foot control panel for injection is also available for some injection systems). Carefully remove the needle from the cell with the joystick.
8. Carefully observe the morphological change and cell behavior immediately after injection. Lower the pressure by increments of 5 hPa if inflation or explosion of cells happens during injection.
9. Incubate the cells for a desired time period and do the follow-up assays (see Note 4).

4. Notes

1. Theoretically, only soluble proteins can be injected efficiently. Some sticky proteins, large proteins, and strongly charged proteins are not suitable for injection, since they might block the injection needle tips.
2. In dividing cell lines, the injected substances are constantly diluted by cell division. Therefore, long incubation after injection of dividing cells is not recommended.
3. Since microinjection is a physical insult to cells, a vehicle control is mandatory to exclude the possibility that injection itself significantly affects cell viability.
4. To ensure effective injection, positive controls and immunochemical staining to detect the injected substances are recommended.

Acknowledgments

This work was supported by the National Program of Basic Research sponsored by the Ministry of Science and Technology of China (2009CB941301), Peking University President Research Grant, Ministry of Education Recruiting Research Grant, and Roche Research Grant.

The authors declare no competing of interest. All authors declare no actual or potential conflicts of interest including any financial, personal, or other relationships with other people or organizations within 3 years of beginning the work submitted that could inappropriately influence (bias) their work.

References

1. Zhang, Y., and Yu, L. C. (2008) Microinjection as a tool of mechanical delivery, *Curr Opin Biotechnol 19*, 506–510.
2. Zhang, Y., and Yu, L. C. (2008) Single-cell microinjection technology in cell biology, *Bioessays 30*, 606–610.
3. Zhang, Y., and LeBlanc, A. (2002) *Microinjection of proteins, peptides and genes in human neuronal apoptosis study.*, Vol. 37, 2 ed., Humana Press, Totowa, New Jersey, .
4. Zhang, Y., Champagne, N., Beitel, L. K., Goodyer, C. G., Trifiro, M., and LeBlanc, A. (2004) Estrogen and androgen protection of human neurons against intracellular amyloid beta1-42 toxicity through heat shock protein 70, *J Neurosci 24*, 5315–5321.
5. Zhang, Y., Goodyer, C., and LeBlanc, A. (2000) Selective and protracted apoptosis in human primary neurons microinjected with active caspase-3, -6, -7, and –8, *J Neurosci 20*, 8384–8389.
6. Zhang, Y., Hong, Y., Bounhar, Y., Blacker, M., Roucou, X., Tounekti, O., Vereker, E., Bowers, W. J., Federoff, H. J., Goodyer, C. G., and LeBlanc, A. (2003) p75 neurotrophin receptor protects primary cultures of human neurons against extracellular amyloid beta peptide cytotoxicity, *J Neurosci 23*, 7385–7394.

7. Zhang, Y., McLaughlin, R., Goodyer, C., and LeBlanc, A. (2002) Selective cytotoxicity of intracellular amyloid beta peptide1-42 through p53 and Bax in cultured primary human neurons, *J Cell Biol 156*, 519–529.
8. Zhang, Y., Tounekti, O., Akerman, B., Goodyer, C. G., and LeBlanc, A. (2001) 17-beta-estradiol induces an inhibitor of active caspases, *J Neurosci 21*, RC176.
9. Kacharmina, J. E., Crino, P. B., and Eberwine, J. (1999) Preparation of cDNA from single cells and subcellular regions, *Methods Enzymol 303*, 3–18.
10. Malter, H. E. (1992) *Early development of micromanipulation. In Micromanipulation of Human Gametes and Embryos*, Raven Press, NY.
11. Wadsworth, P. (1999) Microinjection of mitotic cells, *Methods Cell Biol 61*, 219–231.
12. Leonetti, J. P., Mechti, N., Degols, G., Gagnor, C., and Lebleu, B. (1991) Intracellular distribution of microinjected antisense oligonucleotides, *Proc Natl Acad Sci USA 88*, 2702–2706.
13. Romo, X., Pasten, P., Martinez, S., Soto, X., Lara, P., de Arellano, A. R., Torrejon, M., Montecino, M., Hinrichs, M. V., and Olate, J. (2007) xRic-8 is a GEF for Gsalpha and participates in maintaining meiotic arrest in Xenopus laevis oocytes, *J Cell Physiol.*
14. Gallego, M. A., Ballot, C., Kluza, J., Hajji, N., Martoriati, A., Castera, L., Cuevas, C., Formstecher, P., Joseph, B., Kroemer, G., Bailly, C., and Marchetti, P. (2007) Overcoming chemoresistance of non-small cell lung carcinoma through restoration of an AIF-dependent apoptotic pathway, *Oncogene.*
15. Jarve, A., Muller, J., Kim, I. H., Rohr, K., MacLean, C., Fricker, G., Massing, U., Eberle, F., Dalpke, A., Fischer, R., Trendelenburg, M. F., and Helm, M. (2007) Surveillance of siRNA integrity by FRET imaging, *Nucleic Acids Res 35*, e124.
16. Lin, Y. C., Hsieh, L. C., Kuo, M. W., Yu, J., Kuo, H. H., Lo, W. L., Lin, R. J., Yu, A. L., and Li, W. H. (2007) Human TRIM71 and Its Nematode Homologue Are Targets of let-7 MicroRNA and Its Zebrafish Orthologue Is Essential for Development, *Mol Biol Evol 24*, 2525–2534.
17. Khanam, T., Raabe, C. A., Kiefmann, M., Handel, S., Skryabin, B. V., and Brosius, J. (2007) Can ID Repetitive Elements Serve as Cis-acting Dendritic Targeting Elements? An In Vivo Study, *PLoS ONE 2*, e961.
18. Bossi, E., Fabbrini, M. S., and Ceriotti, A. (2007) Exogenous protein expression in Xenopus oocytes: basic procedures, *Methods Mol Biol 375*, 107–131.
19. Venoux, M., Basbous, J., Berthenet, C., Prigent, C., Fernandez, A., Lamb, N. J., and Rouquier, S. (2007) ASAP is a novel substrate of the oncogenic mitotic kinase Aurora-A : phosphorylation on Ser625 is essential to spindle formation and mitosis, *Hum Mol Genet.*
20. Jaulin, F., Xue, X., Rodriguez-Boulan, E., and Kreitzer, G. (2007) Polarization-dependent selective transport to the apical membrane by KIF5B in MDCK cells, *Dev Cell 13*, 511–522.
21. Derheimer, F. A., O'Hagan, H. M., Krueger, H. M., Hanasoge, S., Paulsen, M. T., and Ljungman, M. (2007) RPA and ATR link transcriptional stress to p53, *Proc Natl Acad Sci USA 104*, 12778–12783.
22. Bostrom, P., Andersson, L., Rutberg, M., Perman, J., Lidberg, U., Johansson, B. R., Fernandez-Rodriguez, J., Ericson, J., Nilsson, T., Boren, J., and Olofsson, S. O. (2007) SNARE proteins mediate fusion between cytosolic lipid droplets and are implicated in insulin sensitivity, *Nat Cell Biol 9*, 1286–1293.
23. Broders-Bondon, F., Chesneau, A., Romero-Oliva, F., Mazabraud, A., Mayor, R., and Thiery, J. P. (2007) Regulation of XSnail2 expression by Rho GTPases, *Dev Dyn 236*, 2555–2566.
24. Woodward, B. J., Campbell, K. H., and Ramsewak, S. S. (2007) A comparison of head-first and tailfirst microinjection of sperm at intracytoplasmic sperm injection, *Fertil Steril.*
25. Yoshida, N., and Perry, A. C. (2007) Piezo-actuated mouse intracytoplasmic sperm injection (ICSI), *Nat Protoc 2*, 296–304.
26. Ehrenberg, M., and McGrath, J. L. (2005) Binding between particles and proteins in extracts: implications for microrheology and toxicity, *Acta Biomater 1*, 305–315.
27. Terns, M. P., and Goldfarb, D. S. (1998) Nuclear transport of RNAs in microinjected Xenopus Oocytes., *Method Cell Biol 53*, 559–589.
28. Fish, M. P., and Groth, A. C. (2007) Creating transgenic Drosophila by microinjecting the site-specific phiC31 integrase mRNA and a transgene-containing donor plasmid., *Nat Protoc 2*, 2325–2331.
29. Anand, R. J., Dai, S., Rippel, C., Leaphart, C., Qureshi, F., Gribar, S. C., Kohler, J. W., Li, J., Beer Stolz, D., Sodhi, C., and Hackam, D. J. (2008) Activated Macrophages Inhibit Enterocyte Gap Junctions via the Release of Nitric Oxide, *Am J Physiol Gastrointest Liver Physiol 294*, G109–119.

30. Green, M., Thorburn, A., Kern, R., and Loewenstein, P. M. (2007) The use of cell microinjection for the in vivo analysis of viral transcriptional regulatory protein domains, *Methods Mol Med 131*, 157–186.
31. Giovannardi, S., Soragna, A., Magagnin, S., and Faravelli, L. (2007) Functional expression of type 1 rat GABA transporter in microinjected Xenopus laevis oocytes, *Methods Mol Biol 375*, 235–255.
32. Hou, J. F., Cui, J., Yu, L. C., and Zhang, Y. (2009) Intracellular amyloid induces impairments on electrophysiological properties of cultured human neurons, *Neurosci Lett 462*, 294–299.
33. Bounhar, Y., Zhang, Y., Goodyer, C. G., and LeBlanc, A. (2001) Prion protein protects human neurons against Bax-mediated apoptosis, *J Biol Chem 276*, 39145–39149.
34. Guo, H., Albrecht, S., Bourdeau, M., Petzke, T., Bergeron, C., and LeBlanc, A. C. (2004) Active caspase-6 and caspase-6-cleaved tau in neuropil threads, neuritic plaques, and neurofibrillary tangles of Alzheimer's disease, *Am J Pathol 165*, 523–531.
35. Guo, H., Petrin, D., Zhang, Y., Bergeron, C., Goodyer, C. G., and LeBlanc, A. C. (2006) Caspase-1 activation of caspase-6 in human apoptotic neurons, *Cell Death Differ 13*, 285–292.
36. Roucou, X., Giannopoulos, P. N., Zhang, Y., Jodoin, J., Goodyer, C. G., and LeBlanc, A. (2005) Cellular prion protein inhibits proapoptotic Bax conformational change in human neurons and in breast carcinoma MCF-7 cells, *Cell Death Differ 12*, 783–795.
37. Roucou, X., Guo, Q., Zhang, Y., Goodyer, C. G., and LeBlanc, A. C. (2003) Cytosolic prion protein is not toxic and protects against Bax-mediated cell death in human primary neurons, *J Biol Chem 278*, 40877–40881.
38. Tounekti, O., Zhang, Y., Klaiman, G., Goodyer, C. G., and LeBlanc, A. (2004) Proteasomal degradation of caspase-6 in 17beta-estradiol-treated neurons, *J Neurochem 89*, 561–568.

Chapter 14

Ultrasonic Manipulation of Single Cells

Martin Wiklund and Björn Önfelt

Abstract

Ultrasonic manipulation has emerged as a simple and powerful tool for trapping, aggregation, and separation of cells. During the last decade, an increasing amount of applications in the microscale format has been demonstrated, of which the most important is acoustophoresis (continuous acoustic cell or particle separation). Traditionally, the technology has proven to be suitable for treatment of high-density cell and particle suspensions, where large cell and particle numbers are handled simultaneously. In this chapter, we describe how ultrasound can be combined with microfluidics and microplates for particle and cell manipulation approaching the single-cell level. We demonstrate different cell handling methods with the purpose to select, trap, aggregate, and position individual cells in microdevices based on multifrequency ultrasonic actuation, and we discuss applications of the technology involving immune cell interaction studies.

Key words: Ultrasonic manipulation, Acoustic trapping, Acoustophoresis, Sono-cage, Single cells

1. Introduction

1.1. Background

Methods for manipulating single cells date back to the early twentieth century when Barber demonstrated how to grasp a cell with suction through a hollow glass micropipette tip (1). Today, this method is still the standard technique for handling and manipulation of single cells, although it requires a skillful operator and can easily damage the cell (2). More recently, methods based on external force fields have emerged as a contactless alternative (3). Two such established techniques—negative dielectrophoresis (nDEP) (4) and laser tweezers (5)—are based on electrical and optical forces, respectively. Their main advantage is the high spatial accuracy defined by, e.g., microelectrodes or a focused laser beam, both having length scales of the order of a single cell or even smaller. Of particular interest for this chapter, however, is ultrasonic cell manipulation, sometimes

Sara Lindström and Helene Andersson-Svahn (eds.), *Single-Cell Analysis: Methods and Protocols*, Methods in Molecular Biology, vol. 853, DOI 10.1007/978-1-61779-567-1_14, © Springer Science+Business Media, LLC 2012

called acoustic trapping or ultrasonic standing wave (USW) manipulation (6). This technique provides a simple, powerful, and possibly more gentle tool for trapping, aggregation, or alignment of particles or cells (7), in particular in microfluidic devices (3, 8). However, in comparison to electrical or optical cell manipulation, ultrasonic manipulation is generally not associated with the high spatial accuracy needed for single-cell handling. Instead, ultrasound is known as an efficient tool for, e.g., high-throughput cell separation ("acoustophoresis") (8) or cell aggregation in mL-volume cell suspensions (9) capable of simultaneous handling of cell numbers ranging from thousands to several millions. Nevertheless, ultrasound has recently been demonstrated to be capable of cell manipulation approaching the single-cell level (10, 11). This chapter focuses on the principles, design, and biocompatibility of devices for ultrasonic manipulation of single cells, and their use for studying immune cell interactions.

1.2. Principles

The principle behind ultrasonic manipulation of small, suspended particles is based on the time-averaged acoustic radiation force obtained from a nonlinear effect in the acoustic pressure field. More than a century ago, Lord Rayleigh described this nonlinear effect as the difference between the average pressure at a surface moving with the sound field and the pressure that would have existed in the fluid of the same mean density at rest (12). This simple definition was not only followed by more rigorous theoretical analyses on the force acting on suspended particles, but also many discussions about the physical interpretation of the phenomenon (13). Today, a generalized model first presented by Gor'kov in 1962 (14) is the most commonly used equation for predicting the acoustic radiation force, **F**, in an arbitrary sound field, p, (15):

$$\mathbf{F} = -\frac{V_\mathrm{p}\beta_\mathrm{f}}{4}\nabla\left(f_1 p^2 - \frac{3}{2k^2} f_2 (\nabla p)^2\right)$$

with

$$f_1 = 1 - \frac{\beta_\mathrm{p}}{\beta_\mathrm{f}} \quad \text{and} \quad f_2 = 2\frac{(\rho_\mathrm{p} - \rho_\mathrm{f})}{2\rho_\mathrm{p} + \rho_\mathrm{f}}.$$

Here, p is the acoustic pressure amplitude, V_p is the volume of the particle, $\beta = 1/(\rho c^2)$ is the compressibility (defined by the density, ρ, and the sound speed, c), $k = \omega/c$ is the wave number, and f_1 and f_2 are the acoustic contrast factors defined by the compressibility β and the density ρ. The index "f" denotes "fluid" and the index "p" denotes "particle." From the equation, we conclude that the radiation force drives suspended particles in a direction parallel with the gradient of the acoustic field and has a direction and magnitude defined by the contrast factors f_1 and f_2. Obviously, steeper gradients result in stronger forces. Now, if we aim for manipulating single cells, it should be preferable to have an acoustic field with gradients of scales approaching the size of the cell (i.e., approximately 10 μm). Two simple methods for creating strong gradients are available: either to focus a propagating acoustic wave or to set up a standing wave by multiple reflections in a resonating chamber. The latter method is by far the most common one (6), resulting in particles or cells driven to the pressure nodes of the USW. Once in the pressure nodes, the particles tend to aggregate into large and flat clusters oriented parallel with the reflecting walls of the chamber and with half-wavelength spacing between the clusters (in chambers with several pressure nodes). The size of these planar clusters can be further decreased by combining the standing wave with a focusing resonator geometry (16, 17). In these studies, acoustic traps were investigated with sizes of the order of a few hundred microns at frequencies up to approximately 10 MHz. Thus, this is the typical size range that defines the spatial accuracy of an acoustic tweezer aimed for single-cell handling. In other words, under these circumstances, it is very difficult to use ultrasound for selectively manipulating a single cell in a high-cell-density bulk suspension. Indeed, one solution would be to decrease the size of the acoustic field gradient by extending the acoustic frequency into the range 10–100 MHz. However, this is difficult to achieve with conventional bulk acoustic wave (BAW) technology, which is typically limited to frequencies lower than ~15 MHz. The reason is the thin size of the piezoelectric plate when increasing the frequency (see Subheading 2.2). On the other hand, manipulation at higher frequencies (typically, up to 150 MHz) has been demonstrated by the use of surface acoustic wave (SAW) technology (18). Although SAW technology is very promising and has potential for single cell manipulation, the biocompatibility of ultrasonic manipulation at these frequencies (~30–150 MHz) has to our knowledge not yet been investigated.

Thus, instead of increasing the spatial accuracy by increasing the frequency, our suggestion is to combine sub-10-MHz acoustic trapping with microfluidic control. Here, microfluidics is used to dispense individual cells into the acoustic traps. In the following sections, we describe how to design and operate an acoustic trap combining microfluidics and acoustics, with the purpose to manipulate individual cells.

2. Materials

2.1. Resonator Design

The traditional design of an ultrasonic resonator aimed for particle manipulation in a microfluidic device is to match the cross-section width (or height) of a fluid channel with half the acoustic wavelength, $\lambda/2$, (or a multiple thereof) (6). For a given channel dimension, L,

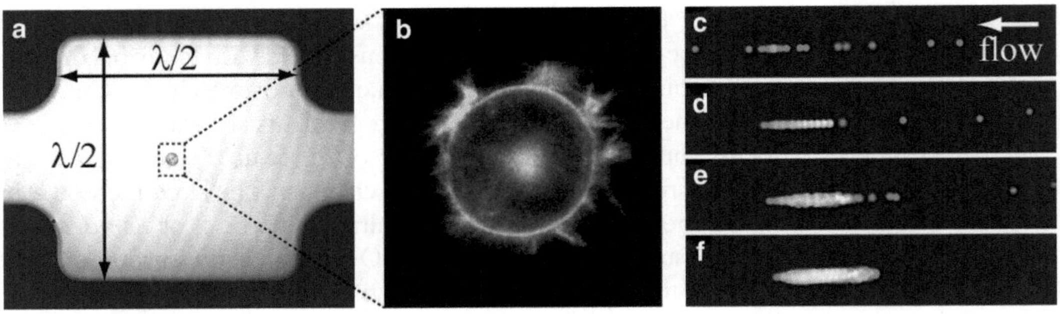

Fig. 1. (**a**) A single trapped human immune cell in a 300 × 300 × 110-μm³ sono-cage. (**b**) High-resolution confocal fluorescence microscopy imaging of the same cell labeled with a membrane probe. (**c–f**) Selected frames from a ~1-min video clip demonstrating trapping and aggregation of 10-μm beads fed into the sono-cage by the use of a continuous fluid flow. Experiment by Björn Önfelt and Otto Manneberg.

the actuation frequency is then given by the relationship $f = c / \lambda = c / 2L$ for a single-node trap (19) or $f = m \times c/2L$ for a multinode trap (where m is an integer) (20). This condition gives a rough guideline of a suitable driving frequency, but in most cases further fine-tuning of the frequency needs to be performed for optimal manipulation efficiency. For single-cell three-dimensional (3-D) manipulation, the strategy is to combine orthogonal half-wave resonances with one $f = c/2L$ condition for each direction along the x, y, and z axes. In practice, this is realized by introducing a rectangular cuboid compartment, called a *sono-cage*, into a microchannel, where individual cells can be trapped and retained close to the center of the cage, see Fig. 1. This compartment typically has larger dimensions than the channel cross-section area. The modes of operation of different sono-cage designs are described in Subheading 3.

In contrast to macroscaled (i.e., mm to cm scaled) ultrasonic manipulation devices (7), resonators integrated into microfluidic chips have proven to be less dependent on the outer device design, such as the dimensions of the solid structures surrounding the microfluidic channel. Macroscaled devices are often designed as multilayer resonators, where the acoustic transducer is part of the resonator (6, 19, 21). In such systems, it is very important to accurately select proper layer dimensions of both the transducer and the supporting structures to the fluid chamber. The reason for this is the coupling of resonances between different layers, which makes it a relatively complex task to predict the optimal driving frequency. Microscaled devices, however, can be actuated from almost an arbitrary coupling point with good performance, as long as the fluid channel is correctly dimensioned relative to the acoustic wavelength. One good example is a method developed by Laurell and coworkers (8), where a large transducer covering the entire bottom surface of the chip is used to excite a standing wave in the fluid channel.

Interestingly, this standing wave is typically selected to have a propagation direction perpendicular to the vibration direction of the transducer element without causing any apparent coupling problem. A related strategy uses Lamb-type plate vibrations of a glass layer in contact with the fluid chamber (20). In another example developed by us (22), a smaller (few mm wide) transducer with a coupling wedge is placed close to a corner of the chip. This method makes it possible to use several transducers operating at many different frequencies, and also to use high-resolution transillumination optical microscopy (where the latter is important for characterizing single cells). A last example mentioned here is to integrate a minimal (~1-mm wide) transducer directly in the fluid channel and use a similar design as the conventional macroscale multilayer resonator design (23). Although this method was originally designed for high-throughput trapping and separation applications (24, 25), it can be used for single-cell manipulation if it is combined with accurate microfluidic control (26).

The standing wave inside the fluid channel is built up by multiple reflections in the supporting channel walls. Thus, these walls act as acoustic mirrors. Two mirror properties need to be considered here: reflectivity and surface roughness. In acoustics, reflectivity, R_I (i.e., relative reflected intensity), is calculated from differences in acoustic impedance, Z, between the medium (fluid) and the reflector (chip material) according to $R_I = \left[(Z_{\text{refl}} - Z_{\text{fluid}}) / (Z_{\text{refl}} + Z_{\text{fluid}})\right]^2$. Approximate values of R_I in common chip materials with water channels are 70% for silicon, 60% for glass, 10% for PMMA, 5% for SU-8, and 4% for PDMS. Therefore, silicon and glass are the most popular materials (8), but plastics have also been employed in a few studies (27, 28). The main reason that plastics work under certain circumstances in BAW transducer systems is due to the fact that the reflectivity from any fluid or solid to air is of the order of 99.9% or higher. Thus, all energy delivered to the chip remains in the transducer—chip system including the fluid channel. Therefore, the material choice is more a question about losses and geometry than reflectivity. Losses can be quantified by the quality factor (Q) of the resonator, which is a measure of the amplification of a vibration in the resonator. Q is also a measure of the sharpness of the resonance frequency according to $Q=f_c/\Delta f$ (where f_c is the center frequency and Δf is the bandwidth). Low losses result in high-Q (frequency specific) resonators while high losses cause the opposite leading to damping and heating. Typically, silicon and glass are high-Q materials while plastics are low-Q materials. When the device is used with optical microscopy, the damping properties of a chip holder may also influence the Q value and frequency response.

The other reflectivity property to be considered is the surface roughness of the acoustic mirrors (i.e., the channel walls). Just like in optics, the roughness of a mirror is related to the scale of the wavelength. However, ultrasonic wavelengths in water at 1–10-MHz

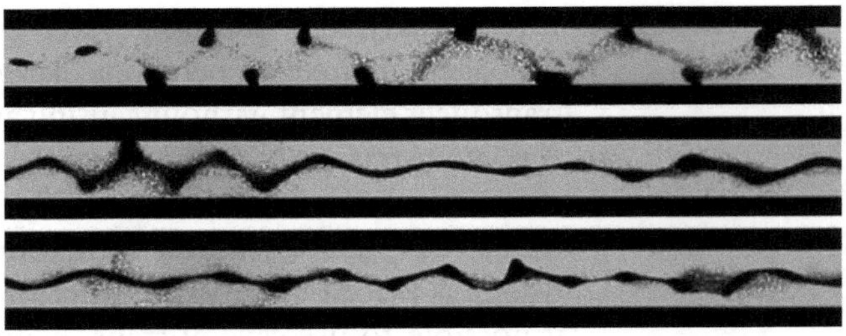

Fig. 2. Demonstration of strong mode coupling when a microfluidic channel is driven in half-wave mode, causing a wavy and striated line of focused particles. Here, the three images show the effect of changing the transducer geometry but for the same actuation frequency. Similar patterns appear when fine-tuning the actuation frequency. Experiment by Ida Iranmanesh.

frequencies are of the order of a few hundred microns. Thus, any roughness with scale 1–2 orders of magnitude lower (i.e., a few microns) is invisible to the ultrasonic wave. A good example demonstrating this is a study by Laurell and coworkers, in which a relatively rough and also asymmetric microchannel etched in glass showed a sufficiently good efficiency in manipulating suspended particles (29).

The last matter to discuss related to resonator design is the problem with unwanted mode coupling in orthogonal directions in a microchannel. For example, if the resonating direction in a rectangular cuboid compartment is along the longest side, there is no risk of any mode coupling in an orthogonal direction as long as the resonator is driven in single-node (half-wave) mode. The obvious reason is that any other direction in the rectangular cuboid then corresponds to less than a half-wave at this driving frequency. However, in acoustophoresis, the resonating direction is typically oriented across the channel width while the longest cuboid side is along the fluid channel. This means that when the half-wave condition is fulfilled across the channel width there is a significant risk of having an $f = m \times c/2L$ condition fulfilled along the channel at the same frequency (where m is an integer). The risk actually increases with increasing aspect ratio (length vs. width) of the channel due to the smaller (relative) frequency step when going from m to $m+1$ for higher values of m. A more theoretical approach to this problem is discussed in ref. 30. In practice, the effect of mode coupling appears as a nonuniform manipulation performance along the channel, often with a wavy and/or striated alignment pattern of focused particles, see Fig. 2. Strategies to avoid or circumvent this problem are discussed in Subheading 3.1.

2.2. Transducer Design

Most transducers used for ultrasonic manipulation of particles are based on a piezoelectric ceramic plate, often referred to as a piezoceramic plate, crystal, or simply a "piezo." A common material is lead zirconate titanate (PZT), which is a ceramic compound of lead, zirconium, and titanate. For particle manipulation purposes, hard PZT materials are preferred, such as the material "Pz-26" from Ferroperm Piezoceramics A/S, Denmark. When applying an AC voltage over the piezo, a mechanical vibration is induced capable of transmitting an ultrasonic wave away from the surface. Typically, a few volts of actuation voltage amplitude corresponds to ~1-nm displacement amplitude of the vibrating piezo surface, which is enough for generating acoustic pressure amplitudes of the order of one atmospheric pressure (atm). Typically, 1–10-atm pressure amplitudes are needed for efficient cell manipulation. This pressure range is also safe from a biocompatibility point of view (see Subheading 3.7).

A naked piezo can act as an efficient transducer itself, but it is often combined with a coupling layer and a protective housing. Most of the devices described in Subheading 3 use wedges as coupling layers for oblique acoustic coupling into the chip (22). The main purpose of a coupling layer is to increase the transmission by impedance matching, but it can also act as a protective layer between the piezo and the chip, as a heat sink, or as a spacer for minimizing near-field effects. Suitable coupling layers for transducers used on microfluidic chips are, e.g., aluminum and titanium for coupling into glass and silicon, respectively (see Note 1). It should also be mentioned that a coupling medium is needed in between the transducer and the chip to fill out all microscopic air gaps present between the two solid surfaces facing each other (see Note 2). Typically, an acoustically transparent fluid is used, such as a water-based gel, glycerol, oil, or glue. The coupling medium should be as thin as possible and assembled in such a way that no air bubble is entrapped in between the transducer and chip. A too thick layer of coupling medium will introduce unwanted losses, in particular if glue or oil is used. In terms of robustness, glue is the most popular choice of coupling medium. On the other hand, gel, glycerol, or oil is a suitable choice if a nonpermanent transducer—chip assembly procedure—is preferred.

Another vital part of a transducer is the backing layer. Traditionally, backing layers are used in medical diagnostic transducers for producing short pulses. However, for particle manipulation purposes, the conventional design is to use air-backed transducers intended for high-Q continuous-wave driving mode. This implies low losses but also very-narrow-frequency bandwidths which makes it more difficult to match the transducer resonance with the resonance of the microfluidic chip. Recently, there has been some interest in designing broadband resonators/transducers for higher flexibility in particle manipulation devices (31).

2.3. Selection and Preparation of Cells for Ultrasonic Manipulation

Several different cell types have been used for ultrasonic manipulation, e.g., yeast (32), plant cells (33), and many different types of mammalian cells (34). Also, various types of bacteria have been manipulated (35). For practical reasons, mostly cell lines which are easy to grow and often quite robust have been used, but there are also examples where primary cells have been used (16, 36, 37). In ultrasonic trapping experiments, it is of particular interest to investigate cells with functions that depend on cell–cell contact. Examples of that are gap junctions formed between adjacent cells (37) trapped by ultrasound. Another example from our group is to study lymphocytes (38), which rely on cell contact and recognition by cell surface receptors to survey other cells for signs of disease (see Note 9). For example, T cells and B cells scan other cells searching for expression of disease-associated antigen, and upon stimulation of the B- or T-cell receptors these cells become activated. Natural killer (NK) cells on the other hand have the ability to directly kill virus-infected or transformed cells. This killing depends on a balance between activating and inhibitory signals mediated by cell-surface receptors on the NK cell and ligands expressed by the target cell. Thus, by using ultrasound to force NK cells and target cells together, investigations of target cell recognition mechanisms can be facilitated; see Subheading 3.5.

Primary lymphocytes can be isolated from blood through negative or positive sorting with magnetic beads or by FACS. The two latter methods have the drawback that the isolated cells are left coated with antibodies and/or magnetic beads.

Typically, the cells need to be fluorescently labeled to (1) separate different cell types from each other and (2) study cell survival or (3) the dynamics of specific proteins inside or on the surface of the cells. Separation of different cell types could be done, e.g., using dyes accumulating in the cell cytoplasm or dyes that bind to cell membranes. For the separation of live and dead cells, several different dyes are available, for example Calcein AM selectively stain the cytoplasm of living cells after hydrolysis of a acetomethyl ester. During the hydrolysis, the dye becomes charged resulting in significantly slower transport across the cell membrane. However, as cells die, the membrane becomes compromised making it possible to study cell death in real time through leakage of Calcein. There are also various dyes for selectively labeling dead cells and many of these are based on dyes leaking into the cell through the compromised plasma membrane labeling the nucleus. Thus, these dyes need to be present in the cell medium if killing is to be studied in real time. However, as several of these dyes are toxic to cells, their usefulness in long-term live-cell imaging is restricted.

For labeling specific proteins, it is possible to use fluorescently labeled antibodies or fab fragments. A restriction with antibodies is that they only label proteins on the surface of live cells and there is always a risk that the presence of the antibody can influence the interaction with other cells by blocking the protein's normal function. It is of course also possible to use molecular biology tools to make cells express proteins that are linked to, e.g., green fluorescent protein (GFP). This technique has revolutionized live-cell imaging as it has made it possible to study the localization and dynamics of many proteins in real time. Drawbacks of using these techniques are that it often results in overexpression of the protein under investigation, raises questions about potential effects of tagging with GFP, and the fact that it is difficult to achieve expression in many types of primary cells.

3. Methods

The following methods review the work by Wiklund and coworkers at the Royal Institute of Technology on the ultrasonic handling of individual cells.

3.1. Prealignment and Transport of Cells

Alignment or focusing of cells in a microchannel is the basic mode of operation in any ultrasonic manipulation device of microfluidic format. If the purpose is to separate cells, either from a suspending medium or from other cells/particles, the term *acoustophoresis* is often used for this mode of operation. The principle is simply to vibrate a chip having a microchannel with constant (and preferably rectangular) cross section at a frequency corresponding to a half-wave across the channel width (see Subheading 2.1). Combining the cell alignment with a continuous laminar flow, this single-frequency and one-dimensional ultrasonic manipulation method is sufficient for achieving a satisfying performance. However, single-cell handling often requires lower cell concentrations and lower flow rates than normally used during high-throughput acoustophoresis. For example, in single-cell applications, it may be important to prevent cell sedimentation and to achieve a more uniform manipulation effect along the microchannel. This would lead to more uniform cell velocities in the (parabolic profiled) fluid flow, and also lower risk of unwanted interaction or attachment of cells to the channel walls. In such cases, the wavy and striated alignment pattern discussed in Subheading 2.1 (and illustrated in Fig. 2) must be eliminated or at least suppressed. A relatively simple and robust approach to solve this problem is to actuate the system with a frequency modulation scheme (39). This method is based on averaging many possible single-resonance frequencies in a microchannel by cycling linear frequency sweeps (i.e., saw-tooth modulation) over an, e.g., ~100-kHz bandwidth around a center frequency of a few MHz. The method works particularly well if the channel is long relative to the width (see Subheading 2.1) and for channels with square-shaped cross sections (i.e., with equal widths and heights). The reason for the latter is that a channel with square-shaped cross section has similar, but in practice never identical, resonance conditions in the two orthogonal directions along the width and height.

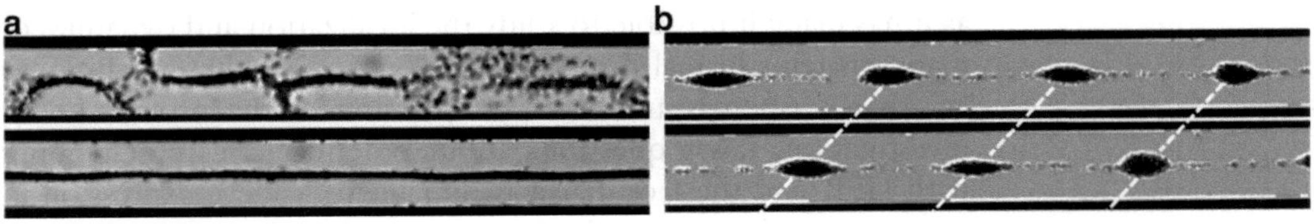

Fig. 3. Demonstration of the effect of fast (**a**) and slow (**b**) frequency-modulation actuation for uniform alignment and flow-free particle transport, respectively. (**a**) The upper panel shows the typical pattern of manipulated particles when the chip is driven at a single frequency (6.9 MHz). The lower panel shows the effect of cycling linear frequency sweeps at the rate 1 kHz, center frequency 6.9 MHz, and bandwidth 100 kHz. (**b**) The procedure in (**a**) is combined with cycling the frequency at the rate 0.5 Hz, center frequency 2.62 MHz, and bandwidth 40 kHz. The panels show two frames from a video clip separated 2 s in time. Experiment by Otto Manneberg.

The cycling rate, i.e., the modulation frequency, should be fast enough to ensure that the cell has no time to move to an area of significantly different force before changing to another resonant mode when changing the "instantaneous" frequency within the linear sweep (40). We have experimentally determined that 1-kHz cycling rate results in a force field corresponding to the average force fields for all single frequencies within the sweep (39). As a result, kHz-frequency-modulation actuation eliminates efficiently the effects of wavy and striated alignment patterns and makes it possible to guide cells along the same fluid streamline close to the center of the microchannel; see Fig. 3a. On the other hand, it is also possible to select a modulation frequency slow enough to ensure that cells do have time to move between different force fields at different frequencies. This is demonstrated in Fig. 3b, where cells are transported acoustically along a microchannel by the use of one fast-frequency-modulation actuation (1-kHz rate) around the center frequency 6.9 MHz (causing uniform cell alignment along the channel), and one slow-frequency-modulation actuation (0.2–0.7-Hz rate) around the center frequency 2.5 MHz (causing periodical cell aggregation and flow-free transport of cells along the channel). For particle transport by slow frequency modulation, the transport speed is controlled electronically by the modulation frequency. The investigated modulation rates are in good agreement with the theoretical cutoff rate of approx. 1 Hz calculated by Glynne-Jones (40).

3.2. Three-Dimensional Trapping of Cells

In Subheading 2.1, the basic principles of three-dimensional cell trapping is presented. The method utilizes an expansion chamber, termed a *sono-cage*, combined with an inlet and an outlet channel (see Fig. 1) for feeding the cage with cells or other particles (10). A slightly different sono-cage design is shown in Fig. 4, where the expansion chamber is formed by two counter-facing cylindrical segments. This sono-cage compartment measures 300 × 600 × 110 μm^3 (width × length × height), and has an inlet and outlet channel with cross-section area 110 × 110 μm^2 (width × height).

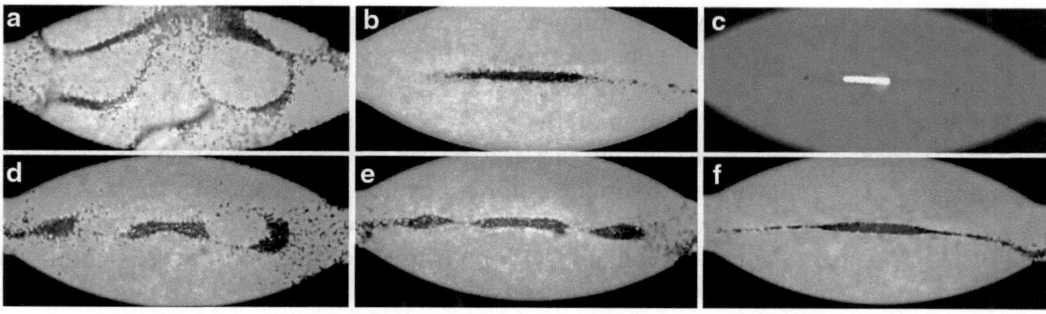

Fig. 4. Demonstration of dual-frequency actuation for three-dimensional ultrasonic manipulation: (**a**) 6.93 MHz/10 V_{p-p}, (**b**) 2.55 MHz/10 V_{p-p}, (**c**) 6.93 MHz/10 V_{p-p} and 2.55 MHz/10 V_{p-p}, (**d**) 6.93 MHz/10 V_{p-p} and 2.55 MHz/2 V_{p-p}, (**e**) 6.93 MHz/10 V_{p-p} and 2.55 MHz/6 V_{p-p}, (**f**) 6.93 MHz/10 V_{p-p} and 2.55 MHz/10 V_{p-p}. All images, except for (**c**), show 5-μm beads; (**c**) shows 10-μm fluorescent beads. Experiment by Otto Manneberg.

Let us now investigate how to perform 3-D trapping of particles or cells by the use of multiple-frequency ultrasonic actuation. In Fig. 4a, b, the trapping pattern of 5-μm polystyrene beads is shown when the sono-cage is excited at 6.93 and 2.55 MHz, respectively. Here, the actuation voltage is 10 V_{p-p} for both frequencies. The higher frequency is used for levitating the beads vertically from the microchannel bottom, and the lower frequency is used for focusing the beads horizontally into a compact aggregate. However, for the levitating frequency (6.93 MHz, Fig. 4a), there is an additional effect creating a complex trapping pattern within the horizontal plane. The reason for this is due to the mode coupling effect discussed in Subheading 2.1. We also note that the focusing frequency (2.55 MHz, Fig. 4b) creates an aggregate of beads on the bottom of the cage chamber. Figure 4c shows the trapping patterns of fluorescent 10-μm beads when the sono-cage is excited with both frequencies (6.93 and 2.55 MHz) simultaneously. Here, the beads form a one-dimensional aggregate, which is trapped, levitated, and positioned close to the center of the sono-cage compartment. Thus, this dual-frequency actuation scheme makes it possible to trap and position the beads three dimensionally in a single point.

In Fig. 4d–f, the effect of gradually increasing one of the actuation voltages, while keeping the other constant, is demonstrated. Here, the 6.93-MHz transducer is excited at 10 V_{p-p} constant voltage while the voltage over the 2.55-MHz transducer is increased from 2 V_{p-p} (Fig. 4d) to 6 V_{p-p} (Fig. 4e) and finally to 10 V_{p-p} (Fig. 4f). Thus, the trapping pattern of dual-frequency actuation in Fig. 4d–f is an amplitude-weighted superposition of the individual single-frequency patterns in Fig. 4a, b. Importantly, the difference between the patterns in Fig. 4b, f is the lack of the horizontal levitation effect in Fig. 4b. Furthermore, the levitation frequency (6.93 MHz) also causes prealignment of the incoming beads, which also enhances the trapping efficiency. Finally, it should be noted that in Fig. 4c, f the transducers are operated identically (i.e., both frequencies and same amplitudes), but with different sizes and concentrations of beads.

3.3. Controlling the Shape of a Trapped Cell Aggregate

An extension of the method presented in Subheading 3.2, in particular in Fig. 4d–f, is to use different amplitude ratios of two actuation frequencies for controlling the shape of a trapped cell or particle aggregate. This is illustrated in Fig. 5, where a constant actuation voltage (10 V_{p-p}) is used for the transducer operating at the levitation frequency (6.81 MHz) while the voltage over the transducer operating at the focusing frequency (2.57 MHz) is varied from 3 to 7 V_{p-p}. The effect is that the shape/dimension as well as the orientation of the aggregate can be controlled electronically; from horizontal 2-D (Fig. 5a), via compact 3-D (Fig. 5b), to vertical 2-D (Fig. 5c). This function can be of interest in studies, where it is important to control the number of neighbors in contact with each cells.

3.4. Selective Trapping of Cells

As mentioned in Subheading 1.1, ultrasonic traps generally do not have the spatial accuracy needed for precise and selective handling of single cells. The main reason is the >100-μm size of the trapping site of a typical device operating in the 1–10-MHz range. Instead, selective cell trapping can be performed if ultrasonic manipulation is combined with accurate fluid control. An example is shown in Fig. 6. Here, a large (multinode) sono-cage is combined with a prealignment channel (41). The prealignment channel can be operated in two different modes, either for single-node alignment at ~2 MHz (half-wave resonance) or for dual-node alignment at ~4 MHz (full-wave resonance). A third frequency (~7 MHz) is used for trapping cells that are injected along the central fluid streamline into the sono-cage element. As a result, 2- and 7-MHz operation leads to continuous trapping and retention of cells in the sono-cage while 4- and 7-MHz operation leads to continuous

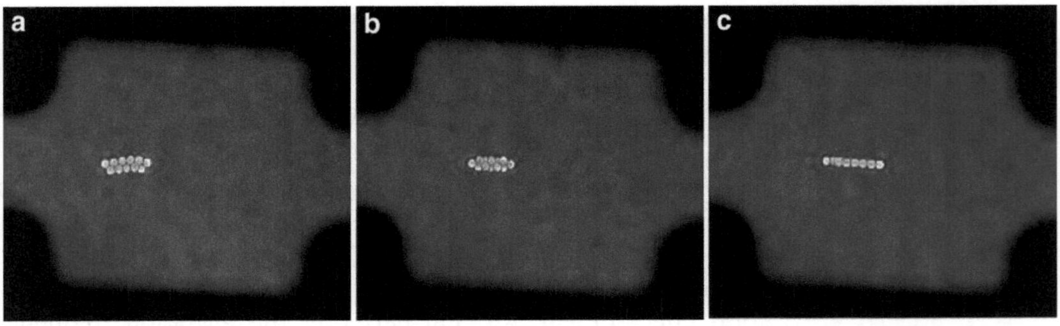

Fig. 5. Demonstration of controlling the shape and position of a bead aggregate: (**a**) 6.81 MHz/10 V_{p-p} and 2.57 MHz/3 V_{p-p} for a horizontal 2-D aggregate. (**b**) 6.81 MHz/10 V_{p-p} and 2.57 MHz/4 V_{p-p} for a 3-D aggregate. (**c**) 6.81 MHz/10 V_{p-p} and 2.57 MHz/7 V_{p-p} for a vertical 2-D aggregate. Experiment by Björn Önfelt and Otto Manneberg.

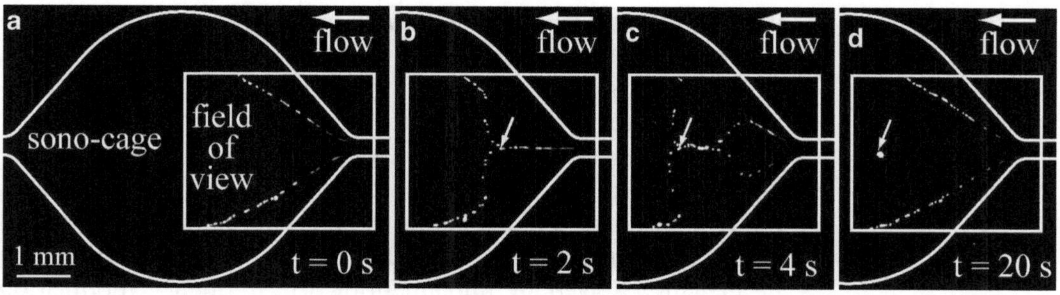

Fig. 6. Frames from a video clip demonstration of selective particle trapping and retention in a fluid flow. The large, rounded curves mark the microchannel boundaries including a 5-mm-wide sono-cage element. The *rectangle marks* the field of view of the microscope for monitoring the paths of 5-μm beads. (**a**) 4- and 7-MHz actuation for particle bypassing. (**b**) 2- and 7-MHz actuation for particle injection. (**c**) 4- and 7-MHz actuation, back to bypassing. (**d**) Final trapping result. Experiment by Jessica Svennebring and Otto Manneberg.

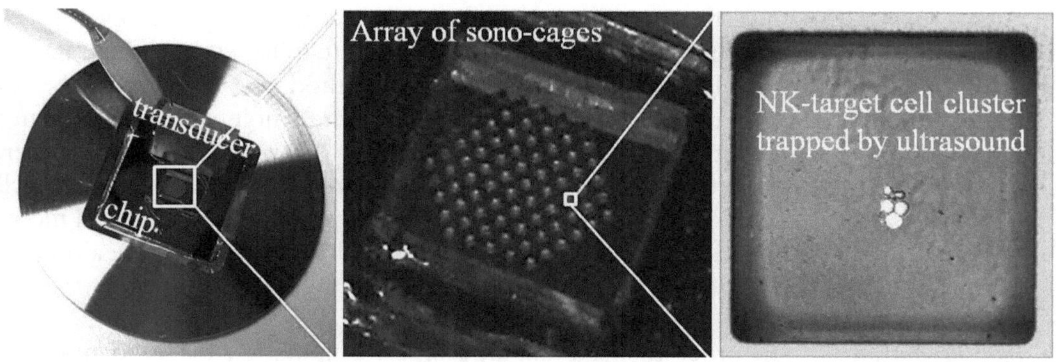

Fig. 7. An array of sono-cages integrated in a multiwell plate, and demonstration of ultrasonic merging of one natural killer (NK) cell and a few target cells. Experiments by Athanasia Christakou.

bypassing of cells through the chip without any trapping. If the actuation is switched from 4 and 7 MHz (Fig. 6a) to 2 and 7 MHz (Fig. 6b) and then back to 4 and 7 MHz (Fig. 6c), the result is injection, trapping, and positioning of a controlled number of cells into the center of the sono-cage element (Fig. 6d). The number of trapped cells can be controlled either manually given that the cell concentration is not too high or by the switching time as long as the cell concentration is constant.

3.5. Parallelized Merging of Cell

A microscale device for ultrasonic manipulation of cells does not necessarily have to be based on a closed microfluidic channel driven in flow-through mode. Another suitable device platform is a multiwell plate or simply a microplate. The basic sono-cage design shown in Fig. 1 (the $300 \times 300 \times 110$-μm^3 rectangular cuboid) can be transferred into the microplate format by just removing the feeding channels and multiply the sono-cages into an array; see Fig. 7. Importantly, the microplate is open and easily accessible, and also free from tubings, valves, and pumps. Therefore, it is a more simple and flexible platform, in particular for the use in long-term cellular studies.

The three-dimensional trapping method described in Subheading 3.2 utilizes two single frequencies operating simultaneously. This simple approach cannot be used in a microplate with an array of sono-cages. The reason is the complex coupling of resonances between different wells, resulting in different trapping patterns and efficiencies for different wells (11). However, this problem can be elegantly solved by implementing the frequency modulation approach described in Subheading 3.1. A similar saw-tooth modulation scheme (i.e., a few MHz center frequency, ~100-kHz bandwidth and 1-kHz sweep rate) implemented in a microplate results in merging, aggregation, and positioning of cells uniformly and simultaneously in all wells (11).

The experiments presented here utilize a simple manual pipetting method for loading cells into the wells: A droplet of cell suspension is placed on top of the plate followed by sedimentation of cells into the wells. With this method, the average number of cells per well can be controlled for the whole plate, but it is not possible to load individual cells in individual wells. However, Andersson-Svahn and coworkers have shown that individual cell loading in a similar multiwell plate as used for ultrasonic cell merging can be performed by the use of a flow cytometer (42), although these wells have significantly larger dimensions.

3.6. High-Resolution Imaging of Trapped Cells

In order to make an ultrasonic manipulation device compatible with high-resolution optical microscopy, some criteria need to be considered in the design process. For microfluidic devices, the most important is to use a material with both good acoustical and good optical properties. An obvious choice is glass, which is optically transparent and has good acoustic reflectivity and low acoustic losses. Therefore, it can also function as an acoustic reflector in a standing-wave resonator. For high-resolution microscopy, the glass layer in the microchip should be of coverslip thickness for optimal image quality (when using microscope objectives that are intended to be used with coverslips). All microchips described in Subheadings 3.1–3.5 are based on bottom glass layers of thickness 200 μm, close to the standard cloverslip thickness. Given this thickness, the glass layer can be used as a quarter-wave acoustic reflector at a frequency of ~6.9 MHz for Pyrex glass. This frequency matches a half-wave chamber of 110 μm in water, which is the choice of microchannel height in the closed devices described in Subheadings 3.1–3.4. Finally, a 1-mm top glass in Pyrex, which is an odd multiple of the quarter-wave thickness of the bottom glass, closes the resonator and provides stability to the chip (see Note 5). Thus, the vertical stack of 200-μm bottom glass, 110-μm silicon and water channel, and 1-mm top glass follows the traditional design of a multilayered acoustic resonator (21) and is used for levitating the cells to the middle of the microchannel (see Subheading 3.2). In addition, this three-layer arrangement is fully transparent and therefore also compatible with any kind of transillumination microscopy technique.

3.7. Biocompatibility

Ultrasonic devices utilize mechanical energy in the form of high-frequency vibrations and pressure fluctuations. This energy may cause damage to biological matters, both at the micro- and macroscale domain. One example of a destructive ultrasound application is high-intensity focused ultrasound (HIFU) used for, e.g., therapy or tissue ablation. In HIFU, heating due to absorption of acoustic energy is the main source of tissue damage. Another phenomenon that may occur in high-intensity ultrasound fields is cavitation, which can be defined as the formation and activity of microbubbles driven into violent oscillation and collapse by the acoustic field. The damage caused by cavitation is partly due to heating effects, but in particular due to mechanical shock waves and liquid jets produced when a bubble implodes. This effect is highly localized into the close vicinity of each cavitation bubble, capable of creating significant microscopic damage to a nearby surface, such as a cell membrane. Today, commercial devices based on acoustic cavitation exist, e.g., for acoustic cell lysis or acoustic membrane poration (sonoporation).

When designing a biocompatible ultrasonic manipulation device, the most important is to ensure that there is no cavitation present and that the temperature is controlled at physiologically correct levels. This strategy is the same as used in diagnostic ultrasound, where the parameters mechanical index (MI) and thermal index (TI) are used for quantifying and monitoring the potential risk of causing damage to tissue or cells as a consequence of cavitation and heating, respectively. In an ultrasonic manipulation device, the temperature can be handled in two different ways. One method is to integrate a cooling system close to the active fluid chamber, e.g., by the use of a cooling water loop or a Peltier cooler. This method is suitable for applications requiring significant powers, such as high-throughput acoustophoresis. Another method is to take advantage of the heat generated by the acoustic actuation and use the temperature increase in a temperature regulation system (43). This method is suitable for medium-power applications, where the acoustic heating corresponds to 1–10°C temperature increases. The latter method can then be combined with additional external heating, e.g., by using a heatable microscope frame as a chip holder. In this way, the temperature can be kept at a constant level (e.g., at 37°C) independently on the magnitude of the acoustic field and for long times (hours to days) (11).

While the temperature is relatively straightforward to monitor and control in an ultrasonic manipulation device, the risk of causing cavitation is unfortunately not. Instead, care must be taken not to use acoustic pressure amplitudes close to the cavitation threshold. This threshold is dependent on many parameters, such as the frequency, type of medium, and actuation mode. As a guideline, Bazou and coworkers have measured the cavitation threshold in a typical ultrasonic manipulation device operating at 1.5 MHz to be 2 MPa (pressure amplitude) (44). This measurement was based on detecting white noise characteristic for cavitation activity by the use of a spectrum analyzer. This threshold can be compared with the typical pressure amplitudes used for efficient particle manipulation, which are in the range 0.1–1 MPa.

If the temperate and CO_2 levels are controlled in a similar way as in a standard cell culture system (typically, at 37°C and 5% CO_2) and if the acoustic pressure amplitude is significantly lower than the cavitation threshold, cells trapped in an ultrasonic manipulation device can be kept viable with retained cellular functions during extended periods in time. In microfluidic devices, additional parameters need to be controlled, such as the biocompatibility of the different chip materials and their surfaces facing the cell sample, as wells as potential shear stress from fluid flows. One of the first cell viability studies performed in a microfluidic device designed for ultrasonic manipulation demonstrated that COS-7 cells acoustically retained in a microchannel at 0.85 MPa for 75 min showed no deviation from normal growth rates when they were returned to the incubator after the ultrasound exposure (45). A more recent study showed that human immune cells were kept viable over 3 days of continuous ultrasound exposure at similar amplitudes (11). In this study, cell proliferation was observed during the exposure. It should be noted, however, that in all these studies cells were trapped and retained in the pressure nodes of an ultrasonic standing wave. Several studies suggest that ultrasonic standing waves are less damaging or stressful for cells than propagating ultrasonic waves (33, 46, 47). One reason could be that pressure fluctuations are more stressful for cells than velocity vibrations. In a pressure node, the pressure fluctuations have a minimum while the velocity vibrations have a maximum. Furthermore, cavitation is more likely to appear in the pressure antinodes than in the pressure nodes if the pressure amplitude is close to the cavitation threshold. Therefore, the acoustic radiation force in an ultrasonic manipulation system actually provides a protective effect on the cells (46).

4. Notes

1. Aluminum is a suitable coupling layer for a PZT transducer aimed for transmitting a wave into a chip made of glass. The reason is that the acoustic impedance of aluminum is in between

the impedance of PZT and glass. For silicon chips, titanium is a suitable coupling layer for the same reason.

2. The layer of liquid coupling medium between the transducer and chip should be as thin as possible and the surfaces should be parallel. The conductive adhesive gel "Tensive" from Parker Laboratories works very well for acoustic coupling and is easily removed with water. Microscope immersion oil works well as coupling medium if it is combined with a holder pressing the transducer against the chip. Since immersion oil is not an adhesive, nor evaporate, it can be used in applications, where the transducer needs to be repositioned during an experiment.
3. Plasma bonding is a suitable method to attach the PDMS frame on the multiwell chip to avoid sample leakage out to the transducer area. A glass lid placed over the frame prevents evaporation. The PDMS frame allows gas exchange between the cell sample in the chip and the environmental chamber, which is important for the long-term biocompatibility of the device.
4. Acetone efficiently dissolves beads that have adhered inside the microchannel. Ethanol followed by water and finally air should be used for washing a chip after a cell experiment.
5. For high-resolution imaging inside a closed microfluidic channel, coverslip glass with thickness ~0.2 mm is suitable as bottom glass of the chip. The other chip layers should be 0.5–1 mm thick to retain stability and robustness of the device. When imaging cells suspended in a microchannel, a water-immersion objective is the best choice. For adherent cells on the bottom of the channel, oil immersion is the best choice.
6. When performing high-resolution imaging over large areas (several mm) based on multiple images, it is important to assemble the chip perfectly parallel with the microscope stage. If available, it is also possible to use automatic focusing.
7. As a cell model, an erythrocyte-mimicking phantom (Orgasol, 5-μm polyamide beads) from Danish Phantom Design is a suitable and inexpensive choice. Polyamide as well as polystyrene beads have acoustic properties relatively similar to cells and respond similarly to the acoustic radiation force.
8. A temperature-controlled system is not only good for the biocompatibility, but also for stabilizing the resonance condition. Since the acoustic wavelength is dependent on the temperature, the manipulation performance may change if the temperature is drifting.
9. Several types of Immune cells are particularly suitable to study with the ultrasonic manipulation technique. One reason is that immune cells grow and function in suspension. Furthermore, immune cells, in particular lymphocytes, have functions that are dependent on cell–cell interactions. Such interactions can be induced by the ultrasonic manipulation technique.

Acknowledgments

The authors thank former and present members of the Ultrasonic manipulation group, Department of Applied Physics, Royal Institute of Technology, Stockholm: Athanasia Christakou, Thomas Frisk, Hans Hertz, Ida Iranmanesh, Linda Johansson, Otto Manneberg, Mathias Ohlin, Jessica Svennebring, and Bruno Vanherberghen for contribution with results and proofreading.

References

1. Barber, M. A. (1904) A new method of isolating microorganisms. *J Kansas Med Sot* 4, 489–494.
2. Fleming, S. D. and King, R. S. (2003) Micromanipulation in Assisted Conception. Cambridge University Press, Cambridge.
3. Nilsson, J., Evander, M., Hammarström, B. and Laurell, T. (2009) Review of cell and particle trapping in microfluidic systems. *Anal Chim Acta* 649, 141–157.
4. Duschl, C., Geggier, P., Jäger, M., Stelzle, M., Müller, T., Schnelle, T. and Fuhr, G. R. (2004) Versatile chip-based tools for the controlled manipulation of microparticles in biology using high frequency electromagnetic fields. In: Andersson, H. and van der Berg, A. Ed. Lab-on-ChipsforCellomics. MicroandNanotechnologies for Life Science. Kluwer Academic Publishers, Dordrecht, The Netherlands.
5. Sott, K., Eriksson, E., Petelenz, E. and Goksör, M. (2008) Optical systems for single cell analyses. *Expert Opin Drug Discov* 3, 1323–1344.
6. Gröschl, M. (1998) Ultrasonic separation of suspended particles – Part I: Fundamentals. *Acustica Acta Acustica* 84, 432–447.
7. Gröschl, M., Burger, W., Handl, B., Doblhoff-Dier, O., Gaida, T. and Schmatz, C. (1998) Ultrasonic separation of suspended particles – Part III: Application in biotechnology. *Acustica Acta Acustica* 84, 815–822.
8. Laurell, T., Petersson, F. and Nilsson, A. (2007) Chip integrated strategies for acoustic separation and manipulation of cells and particles. *Chem Soc Rev* 36, 492–506.
9. Coakley, W. T., Hawkes, J. J., Sobanski, M. A., Cousins, C. M. and Spengler, J. (2000) Analytical scale ultrasonic standing wave manipulation of cells and microparticles. *Ultrasonics* 38, 638–641.
10. Manneberg, O., Vanherberghen, B., Svennebring, J., Hertz, H. M., Önfelt, B. and Wiklund. M. (2008) A three-dimensional ultrasonic cage for characterization of individual cells. *Appl Phys Lett* 93, 063901 (3 pp).
11. Vanherberghen, B., Manneberg, O., Christakou, A., Frisk, T., Ohlin, M., Hertz, H. M., Önfelt, B. and Wiklund, M. (2010) Ultrasound-controlled cell aggregation in a multi-well chip. *Lab Chip* 10, 2727–2732.
12. Lord Rayleigh (1905) On the momentum and pressure of gaseous vibrations, and on the connexion with virial theorem. *Philos Mag* 10, 364–374.
13. Beyer, R. T. (1978) Radiation pressure – the history of a mislabeled tensor. *J. Acoust. Soc. Am.* 63, 1025–1030.
14. Gor'kov, L. P. (1962) On the forces acting on a small particle in an acoustical field in an ideal fluid. *Sov Phys Dokl* 6 773–775.
15. O. Manneberg, Ph.D Thesis, Royal Institute of Technology, Stockholm, 2009.
16. Hertz, H. M. (1995) Standing-wave acoustic trap for nonintrusive positioning of microparticles. *J Appl Phys* 78, 4845–4849.
17. Wiklund, M., Nilsson, S. and Hertz, H. M. (2001) Ultrasonic trapping in capillaries for trace-amount biomedical analysis. *J Appl Phys* 90, 421–426.
18. Wang, Z. and Zhe, J. (2011) Recent advances in particle and droplet manipulation for lab-on-a-chip devices based on surface acoustic waves. *Lab Chip*, in press, DOI: 10.1039/c0lc00527d.
19. Hawkes, J. J., Coakley, W. T., Gröschl, M., Benes, E., Armstrong, S., Tasker, P. J. and Nowotny, H. (2002) Single half-wavelength ultrasonic particle filter: Predictions of the transfer matrix multilayer resonator model and experimental filtration results. *J Acoust Soc Am* 111, 1259–1266.
20. Haake, A. and Dual, J. (2005) Contactless micromanipulation of small particles by an ultrasound field excited by a vibrating body. *J Acoust Soc Am* 117, 2752–2760.

21. Hill, M., Townsend, R. J. and Harris, N. R. (2008) Modelling for the robust design of layered resonators for ultrasonic particle manipulation. *Ultrasonics* 48, 521–528.
22. Manneberg, O., Svennebring, J., Hertz, H. M. and Wiklund, M. (2008) Wedge transducer design for two-dimensional ultrasonic manipulation in a microfluidic chip. *J Micromech Microeng* 18, 095025 (9 pp).
23. Lilliehorn, T., Simu, U., Nilsson, M., Almqvist, M., Stepinski, T., Laurell, T., Nilsson, J. and Johansson, S. (2005) Trapping of microparticles in the near-field of an ultrasonic transducer. *Ultrasonics* 43, 293–303.
24. Evander, M., Johansson, L., Lilliehorn, T., Piskur. J., Lindvall. M., Johansson, S., Almqvist, M., Laurell, T. and Nilsson, J. (2007) Noninvasive acoustic cell trapping in a microfluidic perfusion system for online bioassays. *Anal Chem* 79, 2984–2991.
25. Norris J.V., Evander M., Horsman-Hall K.M., Nilsson J., Laurell T. and Landers J.P. (2009) Acoustic differential extraction for forensic analysis of sexual assault evidence. *Anal Chem* 81, 6089–6095.
26. Johansson, L., Nikolajeff, F., Johansson, S. and Thorslund, S. (2009) On-chip fluorescence-activated cell sorting by an integrated miniaturized ultrasonic transducer. *Anal Chem* 81, 5188–5196.
27. Wiklund, M., Toivonen, J., Tirri, M., Hänninen, P. and Hertz, H. M. (2004) Ultrasonic enrichment of microspheres for ultrasensitive biomedical analysis in confocal laser-scanning fluorescence detection. *J Appl Phys* 96, 1242–1248.
28. González, I., Fernández, L. J., Gómez, T. E., Berganzo, J., Soto, J. L. and Carrato, A. (2010) A polymeric chip for micromanipulation and particle sorting by ultrasounds based on a multilayer configuration. *Sens Act B* 144, 310–317.
29. Evander, M., Lenshof, A., Laurell, T. and Nilsson, J. (2008) Acoustiphoresis in wet-etched glass chips. *Anal Chem* 80, 5178–5185.
30. Barnkob, R., Augustsson, P., Laurell, T. and Bruus, H. (2010) Measuring the local pressure amplitude in microchannel acoustophoresis. *Lab Chip* 10, 563–570.
31. Ohlin, M., Frisk, T., Önfelt, B. and Wiklund, M (2010) Frequency-shift-keying actuation with a damped transducer for ultrasonic particle aggregation in a multi-well chip. *Proc. of USWNet 2010 Conference, Groningen, The Netherlands*, 14–15.
32. Hawkes, J. J. and Coakley, W. T. (1996) A continuous flow ultrasonic cell-filtering method. *Enzyme Microbial Technol* 19, 57–62.
33. Böhm, H., Anthony, P., Davey M.R., Briarty, L.G., Power, J.B., Lowe, K. C., Benes, E. and Gröschl, M. (2000) Viability of plant cell suspensions exposed to homogeneous ultrasonic fields of different energy density and wave type. *Ultrasonics* 38, 629–632.
34. Shirgaonkar, I. Z., Lanthier, S. and Kamen, A. (2004) Acoustic cell filter: a proven cell retention technology for perfusion of animal cell cultures. *Biotechnol Adv* 22, 433–444.
35. Miles, C. A., Morley, M. J., Hudson, W. R. and Mackey, B. M. (1995) Principles of separating micro-organisms from suspensions using ultrasound. *J Appl Bacteriol* 78, 47–54.
36. Cousins, C. M., Holownia, P., Hawkes, J. J., Limaye, M. S., Price, C. P., Keay, P. J. and Coakley, W. T. (2000) Plasma preparation from whole blood using ultrasound. *Ultrasound Med Biol* 26, 881–888.
37. Bazou, D., Dowthwaite, G. P., Khan, I. M., Archer, C. W., Ralphs, J. R. and Coakley, W. T. (2006) Gap junctional intercellular communication and cytoskeletal organization in chondrocytes in suspension in an ultrasound trap. *Mol Memb Biol* 23, 195–205.
38. Guldevall, K., Vanherberghen, B., Frisk, T., Hurtig, J., Christakou, A. E., Manneberg, O., Lindström, S., Andersson-Svahn, H., Wiklund, M. and Önfelt, B. (2010) Imaging Immune Surveillance of Individual Natural Killer Cells Confined in Microwell Arrays. *PLoS One* 10, e15453 (12 pp).
39. Manneberg, O., Vanherberghen, B., Önfelt, B. and Wiklund, M. (2009) Flow-free transport of cells in microchannels by frequency-modulated ultrasound. *Lab Chip* 9, 833–837.
40. Glynne-Jones, P., Boltryk, J. R., Harris, N. R., Cranny, A. W. J. and Hill, M. (2010) Mode-switching: A new technique for electronically varying the agglomeration position in an acoustic particle manipulator. *Ultrasonics* 50, 68–75.
41. Svennebring, J., Manneberg, O., Skafte-Pedersen, P., Bruus, H. and Wiklund, M. (2009) Selective bioparticle retention and characterization in a chip-integrated confocal ultrasonic cavity. *Biotech Bioeng* 103, 323–328.
42. Lindström, S., Larsson, R. and Andersson-Svahn, H. (2008) Towards high-throughput single cell/clone cultivation and analysis. *Electrophor* 29, 1219–1227.
43. Svennebring, J., Manneberg, O. and Wiklund, M. (2007) Temperature regulation during ultrasonic manipulation for long-term cell handling in a microfluidic chip. *J Micromech Microeng* 17, 2469–2474.
44. Bazou, D., Kuznetsova, L. A. and Coakley, W. T. (2005) Physical environment of 2-D animal cell aggregates formed in a short pathlength

ultrasound standing wave trap. *Ultrasound Med Biol* 31, 423–430.
45. Hultström, J., Manneberg, O., Dopf, K., Hertz, H. M. and Wiklund, M. (2007) Proliferation and viability of adherent cells manipulated by standing-wave ultrasound in a microfluidic chip, Ultrasound Med Biol 33, 145–151.
46. Nyborg, W. (2001) Biological effects of ultrasound: Development of safety guidelines. Part II: General review. *Ultrasound Med Biol* 27, 301–333.
47. Chisti, Y. (2003) Sonobioreactors: using ultrasound for enhanced microbial productivity. *Trend Biotechnol* 21, 89–93.

Part III

Reviews on Chosen Subjects Within the Field

Chapter 15

Expanding the Horizons for Single-Cell Applications on Lab-on-a-Chip Devices

Soo Hyeon Kim, Dominique Fourmy, and Teruo Fujii

Abstract

Stochastic events in gene expression, protein synthesis, and metabolite synthesis or degradation lead to cellular heterogeneity essential to life. In a tissue as we see in organs, there is strong heterogeneity among the constituting cells critical to its function. Thus, there exists a strong demand to develop new micro/nanosystems that would enable us to conduct single-cell analysis. This field is rapidly growing, as exemplified below with recent emerging technologies that now reveal sensitive single-cell "omics" analysis. We describe in the review some of the most promising technologies that will certainly transform our view of biology in the near future.

Key words: Single-cell analysis, Genomics, Transcriptomics, Proteomics, Metabolomics, Microsystems, Nanotechnology

1. Introduction

It has been demonstrated that when one looks at the single-cell level, gene expression and protein synthesis are stochastic events (1–4). Therefore, if cells look identical within a population as a whole, they are in fact heterogeneous. Bulk measurements mask any cell-to-cell variation. Capturing this heterogeneity represents an exciting challenge that deeply improves our understanding of life, especially at the tissue level. Recent developments in micro or nanotechnological tools allow "laboratory-on-a-chip" activities now to provide exciting results for single-cell analysis as some of them have been reviewed previously (5–8). In particular, microfluidic devices allow capturing large numbers of single cells in order to analyze their content at that resolution. There is also evidence from a number of experiments that microfuidic devices provide an environment that better mimics in vivo conditions than conventional

Sara Lindström and Helene Andersson-Svahn (eds.), *Single-Cell Analysis: Methods and Protocols*,
Methods in Molecular Biology, vol. 853, DOI 10.1007/978-1-61779-567-1_15, © Springer Science+Business Media, LLC 2012

methods. This might result from high surface area-to-volume ratios and confinement of substances in small volumes, yielding more efficient exchange of substances. It is, therefore, relevant to apply specific stimuli on single cells trapped in the device and analyze their responses. Here, this review focuses on technologies that are capable of capturing single cells and analyzing their constituents in confined microstructures.

2. Lab-on-a-Chip Device for Arraying Single Cells

In order to investigate heterogeneous cellular responses to the environmental stimuli, single cells should be arrayed individually for high-throughput analyses. Microwell array is the most common method for single-cell arraying. Thanks to the MEMS technique, it is possible to fabricate microwells comparable to the size of single cells. Inoue et al. proposed a single-bacterial cell culture system by decreasing the size of microwells (9). By using this system, they found that the length of isolated *Escherichia coli* (*E. coli*) increased at 0.06 μm/min between cell divisions regardless of the chamber volume, and that the cell concentration reached 10^{12} cells/ml under contamination-free conditions. Rettig et al. applied a microwell array technique to high-efficiency single-mammalian cell trapping (10). They examined several experimental parameters, including the diameter and the depth of microwells and seeding time for the optimization study (Fig. 1a). Their results are widely referred to as a technique for single-cell trapping into microwell arrays. Kishi's group used a microwell array for the analysis of cellular responses of individual cells. They analyzed human antigen-specific B cells and produced human monoclonal antibodies (MoAbs) against hepatitis B virus surface antigen (11). Moreover, they developed a microwell array for screening antigen-specific antibody-secreting cells (ASCs) (12). Secreted antibodies diffused out of ASCs in the well and were captured on the chip surfaces, which are coated with antibodies (Fig. 1b). Since the microwell array is open, they could collect selected ASCs for producing MoAbs. This application shows the remarkable merit of the microwell array, that is, accessibility to the target single cells after screening.

Although microwell arrays show good efficiency for single-cell arraying, a drawback of the system lies in the difficulty to change reagents for various chemical stimuli. Figueroa et al. integrated a microfluidic channel into the microwell array platform for high-throughput screening (13). By using calcium imaging, they detected and analyzed odorant responses of about 2,900 olfactory sensory neurons (OSNs) simultaneously. This technique allows both the detection of rare responding OSNs as

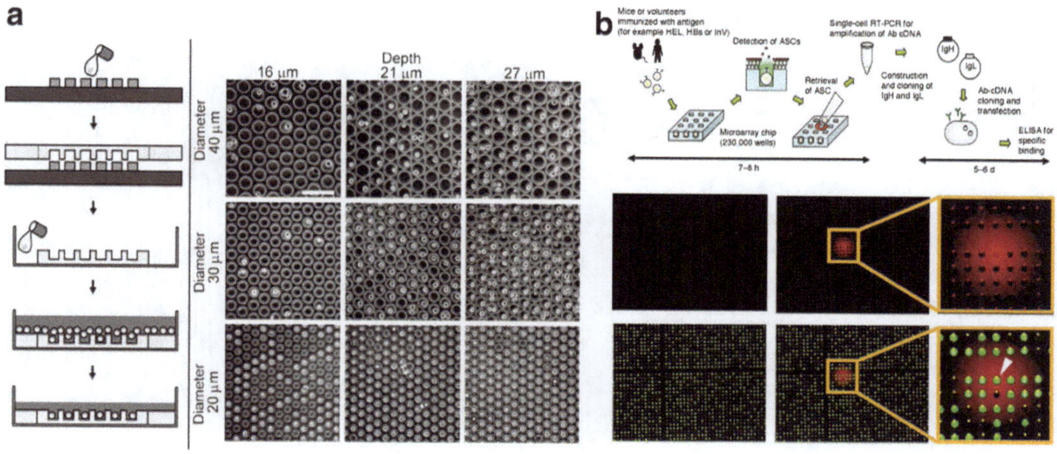

Fig. 1. Microwell array for single-cell analysis. (**a**) Reprinted with permission from ref. 10. Copyright 2005 American Chemical Society. (**b**) Reprinted by permission from Macmillan Publishers Ltd: *Nat Med* (12), copyright 2009.

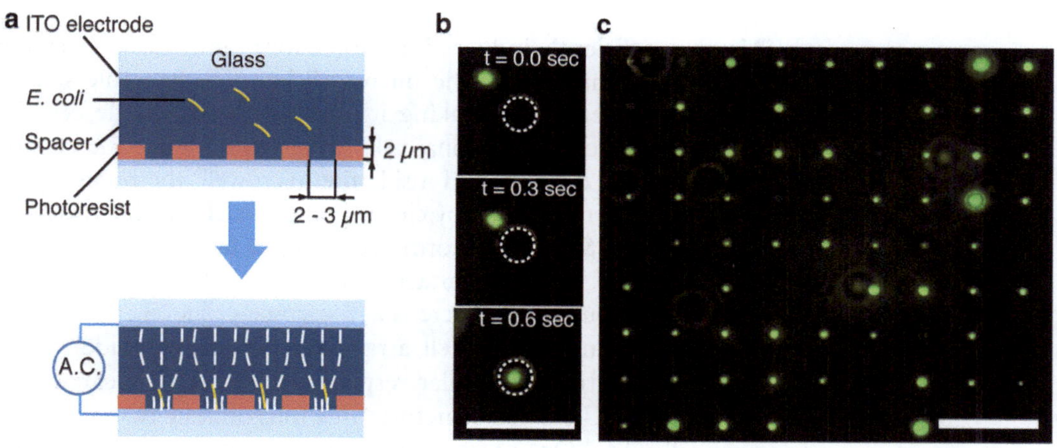

Fig. 2. (**a**) Schematic image of the microwell array for single-bacterial cell trapping and lysing. A highly localized electric field, caused by the through-hole structure, efficiently induces DEP force to trap a single-bacterial cell per microwell. Microwell structure having comparable size to the target cell allows single-cell trapping. (**b**) Time-lapse images of *E. coli* cell trapping. Scale bar is 10 μm. (**c**) Trapped *E. coli* cell array. Scale bar is 20 μm.

well as the identification of OSN populations broadly responsive to various odorants.

The gravity, usually used for trapping single-mammalian cells, cannot attract a single-bacterial cell per microwell. This is because bacterial cells are much smaller in size compared to mammalian cells, typically in the range of femto-liters in volume, and the locomotion by flagella makes handling of bacterial cells difficult. In order to overcome these drawbacks, we have adopted an electrostatic force, dielectrophoresis (DEP), to handle bacterial cells with specificity and stability. A through-hole structure is patterned on an indium tin oxide (ITO)-coated glass slide. The applied electric field is highly deformed and concentrated toward the inside of the microwell structures patterned on the planar electrode (Fig. 2a).

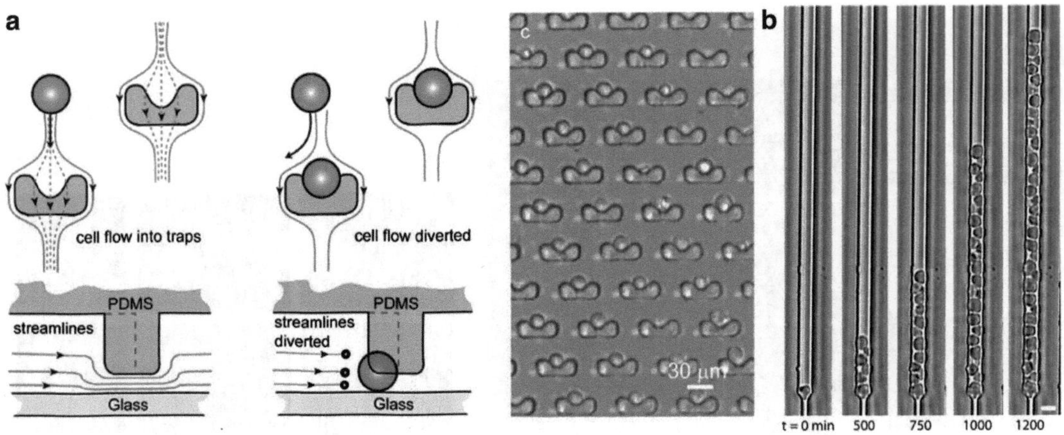

Fig. 3. Hydrodynamic single-cell trapping. (**a**) Reprinted with permission from ref. 14. Copyright 2006 American Chemical Society. (**b**) Reprinted with permission from ref. 16. Copyright 2009 National Academy of Sciences, USA.

This setup efficiently induces DEP force for pulling down bacterial cells into the microwell. The microwell has a comparable size to the target bacterial cell, making it possible to trap single cells by physically excluding additional cells (Fig. 2b, c). Since the electric fields are highly concentrated inside the microwell, the trapped cells can also be effectively lysed by electroporation (EP) in the microwell array platform. Such a platform represents a first step toward the development of highly integrated devices that enable trapping and analysis of a population of bacteria at the single-cell level.

Hydrodynamic single-cell arraying methods are widely used for the observation of cellular response. Carlo et al. created uniform arrays of U-shaped structures in a microfluidic device to trap single cells (14). Once a cell enters into a U-shaped structure, the fraction of streamlines through the structure disappears, leading to the self-closing function of the traps, which facilitates highly efficient trapping (Fig. 3a). The device has advantages on the fluorescent assay and sample preparation, which are manual and intensive operations of incubation, centrifugation, and resuspension in conventional techniques. Well-arranged single-cell isolation arrays facilitate simplified image processing with standard microscopy. The device was applied to determine single-cell enzyme kinetics with fluorogenic substrates for three different cell types (HeLa, 293 T, Jurkat). The hydrodynamic cell trapping system is improved and applied to real-time screening of anticancer drugs against arrays of single cells (15). The platform enables hydrodynamic trapping of cells with relatively low shear stress. Serial observation of ~300 cells in the device shows similar statistical spread to the measurements of single-pass flow cytometer, that is, 15,000–30,000 cells. Rowat et al. developed a microfluidic device for tracking lineages from single cells in parallel (16). Single cells are trapped by mechanical structures in the microfluidic device and are kept to grow and divide in a line (Fig. 3b). By tracking offsprings from single cells, they observe expression levels of the heat-shock protein Hsp12-GFP that fluctuate over time in mother and daughter cells. In contrast, the ribosomal protein Rps8b-GFP shows relatively constant levels of expression over time. Recently, highly improved single-cell trapping performance was realized by Kobel et al. (17). They adapted the Tan and Takeuchi microfluidic trap design optimized for beads (18) to the capture of single-mammalian cells. By the optimization with computational models and flow profiling, the single-cell trapping efficiency is increased up to 97%. With the device, they demonstrate the automated separation of two daughter cells generated upon single-cell division.

3. Lab-on-a-Chip Device for Confinement of Single Cells in an Array

Analysis of biomolecules released from cells provides crucial information on cellular characteristics. As quantities of materials from single cells to be detected are limited, emitted molecules should be physically confined for sensitive detection and to prevent diffusion and cross contamination for a relevant single-cell analysis. Microchambers can be an ideal tool for the confinement of single cells. Cai et al. used microchambers for real-time observation of gene expression in living single cells by confining fluorescent products expelled from the cells (1). By observing the fluorescent products accumulated in the chamber, they show that the expression of β-galactosidase occurs in bursts. Moreover, confinement of antibodies produced by single ASCs was achieved in the chamber (19). Love et al. demonstrated two methods for detecting antibodies. In the first method, they immobilized a secondary antibody at the bottom of the chamber. The secreted antibodies were captured by the secondary antibody, and binding a labeled antigen. In the second one, they immobilized an antigen on the floor of the chamber. Secreted antibodies were captured by the antigen and detected with a labeled secondary antibody. These systems allowed identification, recovery, and clonal expansion of hybridomas for the production of a specific antigen.

Water-in-oil (WO) emulsion droplets can also be used for the confinement of single cells effectively. However, the problem of droplet platforms is low trapping efficiency of single cells, which follows Poisson distribution. Chabert and Viovy increased single-cell encapsulation efficiency to about 70–80% of the injected cell population into drops (20). The device generates relatively smaller droplets compared with target cells (Fig. 4a). Introduced target cells get encapsulated in a droplet of which diameter is larger than that of the empty droplets. The droplets are then sorted depending

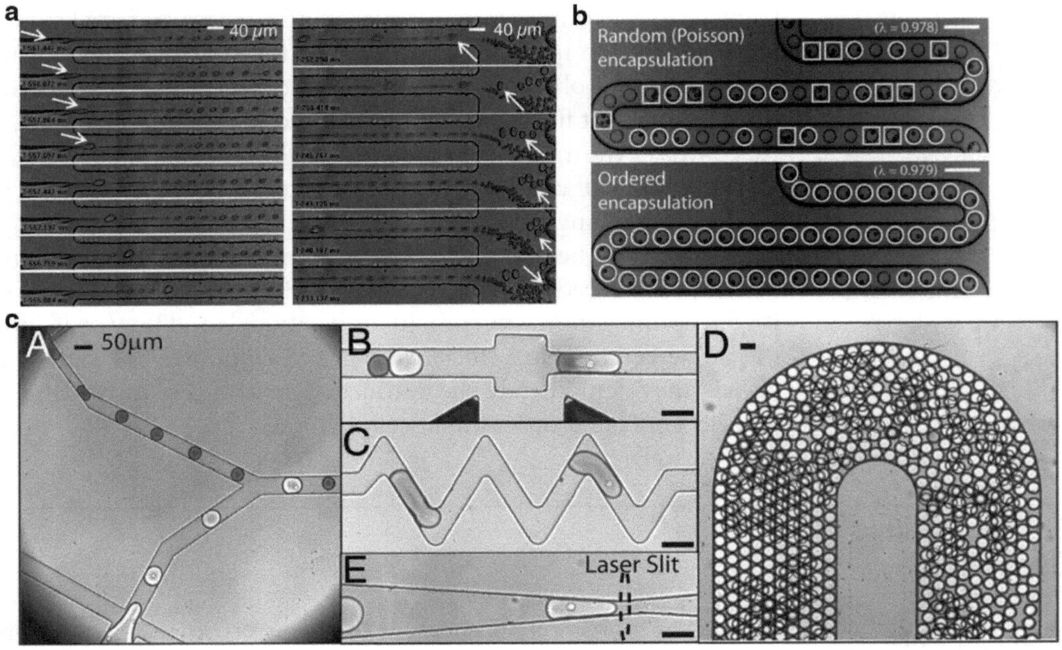

Fig. 4. Single-cell trapping with droplet. (**a**) Reprinted with permission from ref. 20. Copyright 2008 National Academy of Sciences, USA. (**b**) (21) Reproduced by permission of The Royal Society of Chemistry (**c**) Reprinted with permission from ref. 22. Copyright 2009 National Academy of Sciences, USA.

on the size by hydrodynamic mechanisms. The system was applied to direct encapsulation and sorting of cancerous lymphocytes from a whole blood mixture. However, this mechanism has a limitation that sizes of droplets are determined base on the size of the target cell. To overcome this inherent limitation, Edd et al. developed the device having flow-focusing geometry, which focuses and evenly arranges introduced cells before encapsulation (21). Well-arranged cells are introduced into the droplet generation area with the frequency of drop formation (Fig. 4b). The device facilitates controllable loading of single cells into drops, overcoming the intrinsic limitations set by Poisson statistics and ensuring that virtually every drop contains exactly one cell. Brouzes et al. (22) proposed a highly integrated droplet-based microfluidic device for high-throughput cytotoxicity screening of single-mammalian cells (Fig. 4c). The cells in the droplet show high viability over a period of 4 days. Droplet format enables automated generation of optically coded droplet libraries. Droplets from a library are merged with droplets containing target cells for the cytotoxicity screen. The ultimate impact of droplet microfluidic platforms is highly automated manipulation of picoliter volumes, which lead to high-throughput single-cell analyses and combinatorial screening in parallel.

4. Lab-on-a-Chip Device for Analysis of Intracellular Materials of Single Cells

Analysis of intracellular constituents, such as nucleic acids and proteins of single cells, holds great promise for studying diverse biological functions. To get intracellular materials for direct analysis, it is necessary to open the cellular membrane. However, cell membrane disruption is immediately followed by diffusion and dilution of the intracellular material due to the absence of confinement. Since the quantity of the cellular material from single cell is quite small, it is difficult to analyze it when it is diluted. Moreover, cross contamination of intracellular materials among each single cell should be prevented for reliable analyses. Isolation of intracellular materials of individual cells can be achieved by lysing the cell membrane inside tightly enclosed spaces. Lab-on-a-chip devices greatly facilitate analysis of intracellular materials of single cells by enabling the control of ultrasmall volume (pico- to femtoliter).

Since the quantity of intracellular materials of single cell is quite small, much sensitive analysis systems are developed, for some of them, with single-molecule sensitivity. Warren et al. applied single-molecule analysis system to RT-PCR of single cells (23). Prepared cDNA samples from single cells are quantified by using the "digital-counting" method. However, in this technique, isolation of single-cell contents is not automated. Cell sorting was performed using FACS and reverse transcription was done with individual cells in test tubes. Hong et al. automated nucleic acid purification from single cells using microfluidic chips fabricated by multilayer soft lithography (24). The device is equipped with robust and accurate micromechanical valves that prevent any cross contamination or leakage between the steps of the processes (Fig. 5). At the same time, Wu et al. developed an integrated multilayer microfluidic device to manipulate individual cells and chemical reagents, and to accomplish cell lysis and detection of derivatized cellular compounds by using only two valves (25). Such multilayer microfluidic device was used for isolation and genome amplification of individual microbial cells (26), and extraction of total mRNA from individual single cells and synthesis of cDNA on the same device (27, 28). Recently, separation and amplification of homologous copies of each chromosome from a single human metaphase cell were realized in a multilayer microfluidic device (29). Forty-eight chromosomes are released by protease digestion of the cytoplasm from an isolated single cell and randomly separated into 48 chambers for the amplification. Amplified products are corrected and analyzed.

For the isolation of intracellular materials of single cells, Irima et al. (30) used water drops and demonstrated the confinement. Picoliter cell suspension and lysis solution are handled and mixed in the device. They obtained highly concentrated intracellular materials by a limited and stable dilution from a single cell. WO emulsion

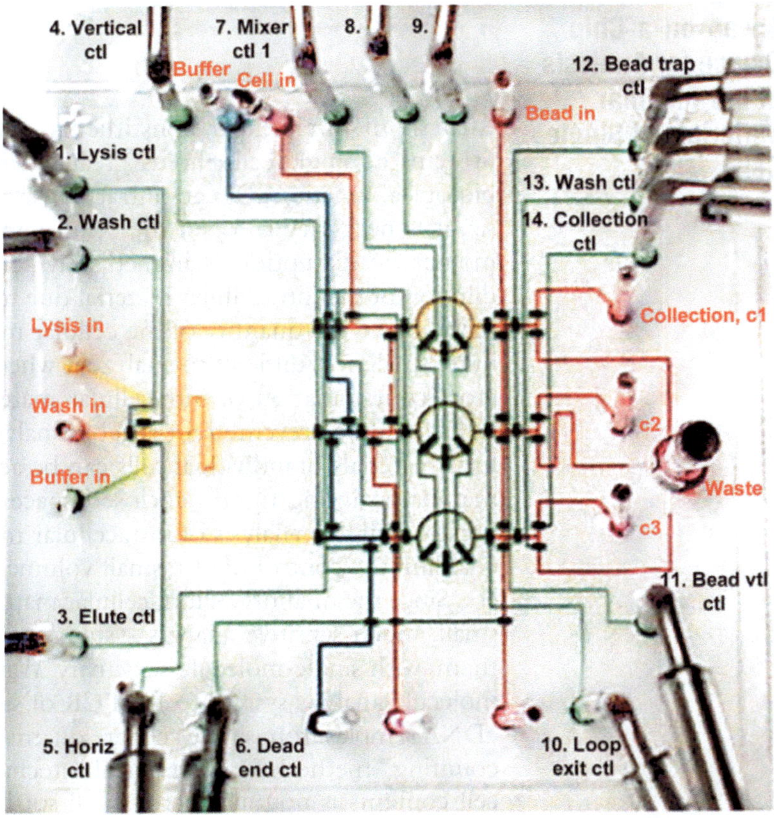

Fig. 5. Microchambers for single-cell analysis. Reprinted by permission from Macmillan Publishers Ltd: *Nat Biotechnol* (24), copyright 2004.

droplets can be a good choice of the reactor for the confinement and analysis of intracellular materials of individual cells. He et al. utilize the droplet format for the intracellular enzymatic reaction (31). Cells are encased into WO droplets by optical trapping and then lysed by laser so that the intracellular materials are confined into the droplet, which prevents dilution and diffusion. They assayed intracellular β-galactosidase by lysing the cells in the droplet with the fluorogenic substrate for the β-galactosidase. However, practical drawbacks of this approach include the low throughput resulting from laser trapping, cell positioning, and lysing (32).

An array format of microchambers for individual cells greatly facilitates high-throughput analysis of intracellular materials in parallel. Isolation of intracellular materials of individual cells can be achieved by lysing the cell membrane inside each of the arrayed microchambers. A volume nearly identical to the targeted cell would make it possible, after lysis, to maintain the cellular components at a concentration close to physiological conditions for subsequent use or analysis. Lee et al. (33) integrated cell lysis function into the microchamber array to analyze intracellular materials of single cells with 10,000 microchambers in parallel.

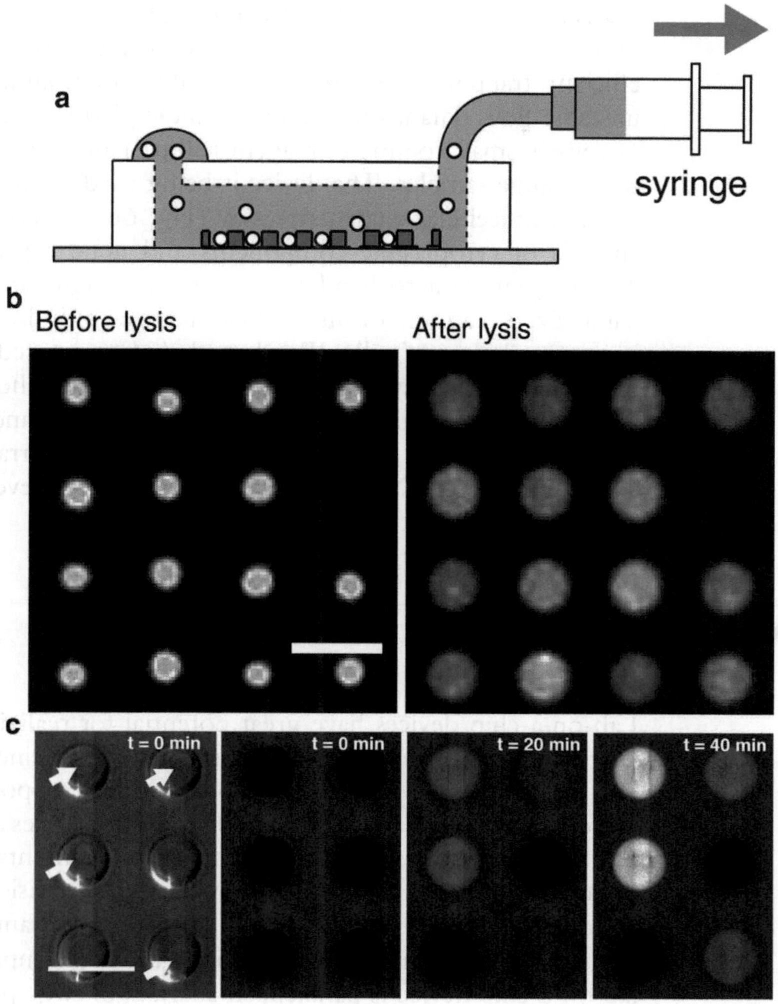

Fig. 6. (**a**) Schematic illustration of our novel microchamber array for single-cell analysis. Target cells are delivered by the continuous flow and trapped by DEP. (**b**) Fluorescence image of trapped cells before and after lysis. Scale bar is 50 μm. (**c**) Enzymatic assay of intracellular-β-galactosidase at the single-cell level. Trapped single cells (*white arrows* indicate trapped cells) are lysed with FDG, fluorogenic substrate of intracellular β-galactosidase. The fluorescence intensity of the microchambers gradually increased with time after cell lysis at $t = 0$ min. The brightness in the array is not homogeneously distributed and likely corresponds to a heterogeneity on the number of galactosidase molecules in individual cells. Scale bar is 50 μm.

Single cells are trapped into a microwell array by gravity. Microwells are closed to form microchambers with an electrode-patterned glass substrate. Trapped single cells are lysed in the tightly enclosed microchambers by electroporation. They investigated the enzymatic activity in single cells by lysing cells with a fluorogenic substrate of an intracellular enzyme. However, their passive trapping method leads to leakage of cell during subsequent microchamber closure process (33). Moreover, the device needs to be dipped into a Petri dish to apply the reagent for analysis, which causes large dead volumes of the reagents. We have developed a novel device to prevent such drawbacks by integrating DEP trapping into a microfluidic platform (Fig. 6a). Interdigitated ITO electrodes, patterned at the bottom of microchambers, attract a single cell per microchamber by inducing DEP force, which provides fast, active, and highly efficient trapping but also allows stable positioning even with unstable flow. Thanks to the microfluidic platform, it requires an extremely small volume of reagent (20 μL) and short time (30 s) to exchange samples. The device is being used for demonstration of the intracellular-enzymatic assay (Fig. 6b, c) and quantitative analysis of cytoplasmic components and gene expression level. The integrated microchamber array provides high-throughput and parallelized readouts of intracellular analytes of individual cells in large population of cells. Wood et al. (34) proposed an agarose microchamber array for single-cell gel electrophoresis. They trapped single cells into an agarose microwell array and performed gel electrophoresis directly using the microwell array structure. They investigate DNA damage at the single-cell level in parallel with fully automated analysis.

5. Conclusion

Lab-on-a-chip devices have great potential for realizing the concept of point-of-care testing for personalized medicine and clinical applications. Now, the device is opening new opportunities for performing an increasing number of "omics" analyses at the single-cell level. Besides direct analysis of single-cell contents for genomics, transcriptomics, and metabolomics, we can envision the use of lab-on-a-chip device for studying intracellular mechanisms, such as cellular pathways and networks. One of the great impacts of using lab-on-a-chip device is parallelized analyses of large population of single cells, giving a great view on the distribution of cell responses, cell characteristics, molecular interactions, intracellular contents, etc. Moreover, lab-on-a-chip devices facilitate the highly sensitive detection and analysis of biological functions by concentrating biomolecules with the ultrasmall volume of reactors comparable in size to the target cell. Especially, microchamber and droplet formats allow analysis of individual cell contents without dilution, diffusion, or cross contamination in parallel. So far, the lab-on-a-chip devices largely rely on light-based detection methods. Therefore, innovative methods, as introduced here, for the analysis of the desirable cellular constituents or biological phenomena are highly required to overcome this limitation. The lab-on-a-chip device with the innovative detection methods realizes advanced biological analyses for precise understanding of genetic heterogeneity on biological functions in a population of cells. The approaches should be of interest for many biological studies and broaden the scope of the biological analyses by serving unique concepts. From this perspective, the future of lab-on-a-chip device for single-cell analysis is promising, and developments going forward will undoubtedly enhance our ability to understand biological functions.

References

1. Cai *et al.* 2006. Stochastic protein expression in individual cells at the single molecule level. *Nature* 440 (7082):358–362. doi:nature04599 (pii) 10.1038/nature04599.
2. Rosenfeld *et al.* 2005. Gene regulation at the single-cell level. *Science* 307 (5717):1962–1965. doi:307/5717/1962 (pii) 10.1126/science.1106914.
3. Weinberger *et al.* 2005. Stochastic gene expression in a lentiviral positive-feedback loop: HIV-1 Tat fluctuations drive phenotypic diversity. *Cell* 122(2):169–182.doi:S0092-8674(05)00549-0 (pii) 10.1016/j.cell.2005.06.006.
4. Suter *et al.* 2011. Mammalian genes are transcribed with widely different bursting kinetics. *Science* 332 (6028):472–474. doi:science.1198817 (pii) 10.1126/science.1198817.
5. Andersson *et al.* 2004. Microtechnologies and nanotechnologies for single-cell analysis. *Curr Opin Biotechnol* 15 (1):44–49. doi:10.1016/j.copbio.2004.01.004 S0958166904000060 (pii).
6. Wang *et al.* 2010. Single cell analysis: the new frontier in 'omics'. *Trends Biotechnol* 28 (6):281–290. doi:S0167-7799(10)00050-8 (pii) 10.1016/j.tibtech.2010.03.002.
7. Lindstrom *et al.* 2010. Overview of single-cell analyses: microdevices and applications. *Lab Chip* 10 (24):3363–3372. doi:10.1039/c0lc00150c.
8. Kalisky *et al.* 2011. Single-cell genomics. *Nat Methods* 8 (4):311–314. doi:nmeth0411-311 (pii) 10.1038/nmeth0411–311.
9. Inoue *et al.* 2001. On-chip culture system for observation of isolated individual cells. *Lab Chip* 1 (1):50–55. doi:10.1039/b103931h.
10. Rettig *et al.* 2005. Large-scale single-cell trapping and imaging using microwell arrays. *Anal Chem* 77 (17):5628–5634. doi:10.1021/ac0505977.
11. Tokimitsu *et al.* 2007. Single lymphocyte analysis with a microwell array chip. *Cytometry A* 71 (12):1003–1010. doi:10.1002/cyto.a.20478.
12. Jin *et al.* 2009. A rapid and efficient single-cell manipulation method for screening antigen-specific antibody-secreting cells from human peripheral blood. *Nat Med* 15 (9):1088–1092. doi:nm.1966 (pii) 10.1038/nm.1966.
13. Figueroa *et al.* 2010. Large-scale investigation of the olfactory receptor space using a microfluidic microwell array. *Lab Chip* 10 (9):1120–1127. doi:10.1039/b920585c.
14. Di Carlo *et al.* 2006. Single-cell enzyme concentrations, kinetics, and inhibition analysis using high-density hydrodynamic cell isolation arrays. *Anal Chem* 78 (14):4925–4930.
15. Wlodkowic *et al.* 2009. Microfluidic single-cell array cytometry for the analysis of tumor apoptosis. *Anal Chem* 81 (13):5517–5523.
16. Rowat *et al.* 2009. Tracking lineages of single cells in lines using a microfluidic device. *Proc Natl Acad Sci USA* 106 (43):18149–18154.
17. Kobel *et al.* 2010. Optimization of microfluidic single cell trapping for long-term on-chip culture. *Lab Chip* 10 (7):857–863. doi:10.1039/b918055a.
18. Tan *et al.* 2007. A trap-and-release integrated microfluidic system for dynamic microarray applications. *Proc Natl Acad Sci USA* 104 (4):1146–1151.
19. Love *et al.* 2006. A microengraving method for rapid selection of single cells producing antigen-specific antibodies. *Nat Biotechnol* 24 (6):703–707. doi:nbt1210 (pii) 10.1038/nbt1210.
20. Chabert *et al.* 2008. Microfluidic high-throughput encapsulation and hydrodynamic self-sorting of single cells. *Proc Natl Acad Sci USA* 105 (9):3191–3196. doi:0708321105 (pii) 10.1073/pnas.0708321105.
21. Edd *et al.* 2008. Controlled encapsulation of single-cells into monodisperse picolitre drops. *Lab Chip* 8 (8):1262–1264. doi:10.1039/b805456h.
22. Brouzes *et al.* 2009. Droplet microfluidic technology for single-cell high-throughput screening. *Proc Natl Acad Sci USA* 106 (34):14195–14200. doi:0903542106 (pii) 10.1073/pnas.0903542106.
23. Warren *et al.* 2006. Transcription factor profiling in individual hematopoietic progenitors by digital RT-PCR. *Proc Natl Acad Sci USA* 103 (47):17807–17812. doi:0608512103 (pii) 10.1073/pnas.0608512103.
24. Hong *et al.* 2004. A nanoliter-scale nucleic acid processor with parallel architecture. *Nat Biotechnol* 22 (4):435–439.

25. Wu *et al.* 2004. Chemical cytometry on a picoliter-scale integrated microfluidic chip. *Proc Natl Acad Sci USA* 101 (35):12809–12813. doi:10.1073/pnas.0405299101 0405299101 (pii).
26. Marcy *et al.* 2007. Dissecting biological "dark matter" with single-cell genetic analysis of rare and uncultivated TM7 microbes from the human mouth. *Proc Natl Acad Sci USA* 104 (29):11889–11894. doi:0704662104 (pii) 10.1073/pnas.0704662104.
27. Zhong *et al.* 2008. A microfluidic processor for gene expression profiling of single human embryonic stem cells. *Lab Chip* 8 (1):68–74. doi:10.1039/b712116d.
28. Bontoux *et al.* 2008. Integrating whole transcriptome assays on a lab-on-a-chip for single cell gene profiling. *Lab Chip* 8 (3):443–450. doi:10.1039/b716543a.
29. Fan *et al.* 2011. Whole-genome molecular haplotyping of single cells. *Nat Biotechnol* 29 (1):51–57. doi:nbt.1739 (pii) 10.1038/nbt.1739.
30. Irimia *et al.* 2004. Single-cell chemical lysis in picoliter-scale closed volumes using a microfabricated device. *Anal Chem* 76 (20):6137–6143. doi:10.1021/ac0497508.
31. He *et al.* 2005. Selective encapsulation of single cells and subcellular organelles into picoliter- and femtoliter-volume droplets. *Anal Chem* 77 (6):1539–1544. doi:10.1021/ac0480850.
32. Sims *et al.* 2007. Analysis of single mammalian cells on-chip. *Lab Chip* 7 (4):423–440.
33. Lee *et al.* 2010. Large-scale arrays of picolitre chambers for single-cell analysis of large cell populations. *Lab Chip* 10 (21):2952–2958. doi:10.1039/c0lc00139b.
34. Wood *et al.* 2010. Single cell trapping and DNA damage analysis using microwell arrays. *Proc Natl Acad Sci USA* 107 (22):10008–10013. doi:1004056107 (pii) 10.1073/pnas.1004056107.

Chapter 16

Analytical Technologies for Integrated Single-Cell Analysis of Human Immune Responses

Ayça Yalçın, Yvonne J. Yamanaka, and J. Christopher Love

Abstract

The immune system is a network of cells in which the constitutive members interact through dense and sometimes overlapping connections. The extreme complexity of this network poses a significant challenge for monitoring pathological conditions (e.g., food allergies, autoimmunity, and other chronic inflammatory diseases) and for discovering robust signatures of immunological responses that correlate with or predict the efficacy of interventions. The diversity among immune cells found in clinical samples (variations in cellular functions, lineages, and clonotypic breadth) requires approaches for monitoring immune responses with single-cell resolution.

In this chapter, we present an engineering approach for integrated single-cell analysis that uses interchangeable modular operations to provide a comprehensive characterization of the phenotypic, functional, and genetic variations for individual cells. We focus on the use of microfabricated devices to isolate and interrogate single cells, and on the analytical components that enable subsequent detection, correlation, and interpretation of multidimensional sets of data. We discuss specific challenges and opportunities in the realization of this concept, and review two examples where it has been implemented. The presented approach should provide a basis for the design and implementation of nonconventional bioanalytical processes for studying specific responses of an immune system.

Key words: Single-cell analysis, Lab-on-a-chip, Micro total analysis, Microengraving, Human immune response, Multidimensional analysis, Single-cell isolation

1. Introduction

Understanding how an immune response develops, and whether it is successful, requires deciphering the relationships that map biological inputs to a person (e.g., pathogens, vaccines, other therapeutic interventions) with the corresponding clinical outcomes, such as infections or protective immunity (Fig. 1). This task is not trivial: The immune system is an extremely complex network of cells with

Sara Lindström and Helene Andersson-Svahn (eds.), *Single-Cell Analysis: Methods and Protocols*,
Methods in Molecular Biology, vol. 853, DOI 10.1007/978-1-61779-567-1_16, © Springer Science+Business Media, LLC 2012

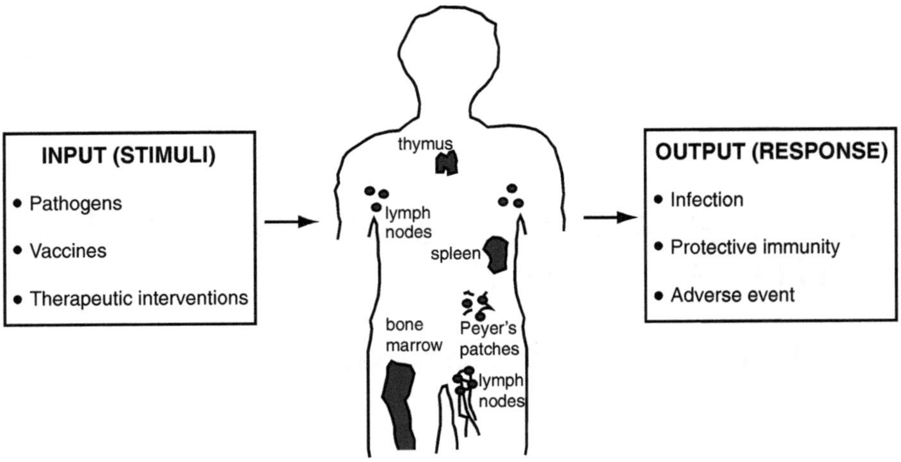

Fig. 1. The immune system as a transfer function for clinical status. When a human is subjected to defined inputs, such as pathogens or vaccines, the highly complex and distributed immune system processes these inputs and responds to them. The response determines to a large extent clinical outcomes, such as an infection or protective immunity. Understanding the pathways and interactions that confer positive and negative outcomes is essential for the rational design of effective therapeutic interventions.

vast numbers of interactions among them. Some estimates suggest that the density of the immune network is comparable to that of the World Wide Web and social networks (1). A significant challenge, therefore, for monitoring pathological conditions, and the efficacy of interventions, is how to determine robust signatures of immunological responses that correlate with or predict either productive or ineffective outcomes. For some diseases, certain measures are strongly associated with the progression of the disease. For example, the number of circulating CD4 T cells is a robust predictor for the development of AIDS following infection by HIV (2). Many clinical conditions, however, including food allergies, autoimmunity, and other chronic inflammatory diseases, are more difficult to assess reliably, and new tools for immune monitoring are needed.

Immune monitoring relies on evaluating clinical samples obtained from affected individuals to measure the molecules from biological fluids (e.g., serum), as well as the types and functions of the immune cells present. An intrinsic challenge in this task results from the limited size and types of samples that can be attained in clinical studies. Samples from large-scale clinical trials and from certain populations, such as pediatrics, are typically limited in volume to a few milliliters of blood containing ~10^6–10^7 lymphocytes. Biopsies of affected tissues may also be obtained in some studies, but the numbers of cells available are even more limited, more expensive, and riskier to obtain than other types of samples. A typical "pinch" biopsy may only yield 10^4–10^5 cells, or enough cells to use in approximately one conventional biological assay for assessing the phenotype or function of the cells in the sample. This constraint significantly restricts the breadth of knowledge available from each sample (3).

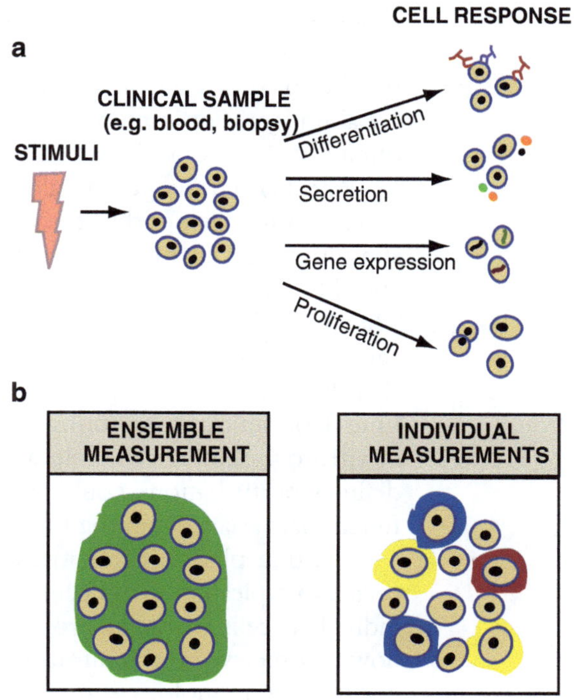

Fig. 2. Cellular analysis is critical for assessing breadth and magnitude of immune responses. (**a**) Cells respond to external stimuli in manifold ways; some of these responses include differentiation of cell types (surface marker change) and dynamic functions (e.g., protein secretion, gene expression, proliferation). (**b**) Measurements of immune responses can involve either average determinations of state from an ensemble of cells (*left*) or resolve contributions from individual cells (*right*).

Immune monitoring must resolve the evolution of an immune response that involves the activation and differentiation of many different subsets of cells, including dendritic cells, macrophages, T cells, and B cells among others. These cells exhibit a wide range of functions and states of differentiation, and in the case of T cells and B cells, significant genetic variations. It is difficult, therefore, to assess the fine nuances of these contributors efficiently. Average measures of the heterogeneity within these subsets of cells can mask the roles that different cell types, especially rare cells or subsets, play in raising an immune response (Fig. 2). To determine differentiating variations in cellular functions, lineages, and clonotypic breadth, it is essential to characterize the diversity in clinical samples of immune cells with single-cell resolution.

1.1. Challenges for Single-Cell Analytical Technologies Used in Immune Monitoring

Existing analytical platforms used to evaluate cellular functions and states of differentiation in vitro have several inherent limitations, whether by design or by circumstance. Conventional bioanalytical systems are intentionally implemented to evaluate one set of related parameters (e.g., gene expression or proteins secreted) without yielding information about other traits (e.g., lineage, genetic variance).

The criteria for designing such systems has tended not to include an objective for conserving sample since in many instances the number of cells available to study from a cell line or animal model are abundant. For immune monitoring in humans, this situation is often not the case as described above.

Another challenge for all bioanalytical technologies is the dynamic variability of the cells. Environmental changes or imposed stimuli induce changes in the biological state of individual cells, and different cells may respond differently to the same input (4). This unpredictability in the evolution of states among individual cells contrasts the nature of conventional analytical measurements that determine an average distribution of states at a fixed point in time. Continuous monitoring of the dynamics of a cell's response in vitro over biologically relevant timescales (minutes to days) to defined perturbations could enhance our understanding of the functional variances among immune cells.

Existing platforms for measuring the responses of individual cells also typically rely on cells isolated by limiting dilution, or on individual cells extracted from a bulk population. For example, flow cytometry—a ubiquitous single-cell analytical platform used in clinical immunology—measures the distributions of populations of single cells after activation or cultivation of a bulk population. The methodology employed in these assays makes it difficult to understand the types and extent of cell-to-cell interactions, and averages over many possible combinations of interacting networks. Developing systems-level knowledge of how a network of immune cells interacts would benefit from technologies that allow precise and defined control over interactions among cells.

Most common assays used to assess the functional variations among immune cells are destructive by their nature. Intracellular cytokine staining (ICS) is a method used with flow cytometers to measure cytokines produced by cells, but the method requires fixing and permeabilizing the cells (5). Another technique, enzyme-linked immunosorbent spot (ELISpot), measures secreted cytokines, but the cells are lost during the process (6). This limitation hinders the extent to which ex vivo functional variations among cells can be related to genetic or clonotypic variations since individual responders are lost. Nondestructive methods would improve the ability to correlate phenotypic and genotypic differences among subsets of T and B cells, and approaches to recover individual viable cells would benefit efforts to relate gene expression to functional variations.

The study of immunological mechanisms using existing technologies has also emphasized a reductionist's view on interpreting the complexity of a multifaceted system. The incorporation of strategies from systems biology has begun to provide a more quantitative understanding of how immune responses evolve, particularly in the context of vaccinations (7, 8). For functional systems immunology, single-cell measurements would help define statistical network models that may improve the knowledge of the specific mechanisms that govern the human response to a disease.

Together, the existing constraints of analytical systems for immune monitoring suggest that new technologies are needed to (1) facilitate the isolation of single cells, (2) identify phenotypical and functional variations, (3) define specific cell–cell interactions, (4) provide the opportunity for multidimensional and time-efficient analysis, and (5) enable quantitative extraction of information. The often-limited size of the clinical samples obtained from patients, and the relative scarcity of cellular events of interest, should also be considered in the fundamental design of new assays, as well as capabilities for increasing throughput and sensitivity to promote their adoption into routine clinical research.

1.2. Scope of this Chapter

This chapter focuses on the design and implementation of integrated systems that use microfabricated devices to isolate and interrogate individual cells, and on the various analytical components that can enable the efficient detection, correlation, and interpretation of multiple types of biologically significant information from such systems. Specifically, we (1) emphasize a new strategy for integrated single-cell analysis (iSCA) that provides a conceptual framework for enabling modularly designed bioanalytical processes that overcome some of the limitations and challenges in conventional analytical technologies; (2) identify the essential components of an integrated analysis system; (3) provide an understanding of how various technologies can be adopted in such systems and how extracted information is linked to individual cells; and (4) suggest the necessary attributes of technologies required to expand the capabilities of iSCA further for the study of human disease.

2. Integrated Single-Cell Analysis

The advancement of techniques to fabricate micro- and nanometer-scale structures over the last two decades has accelerated the maturation of lab-on-a-chip (LOC), or micro total analysis systems (μTAS), to address the challenges for single-cell analysis in immune monitoring (9, 10). LOC systems employ a microfabricated device, or chip, on which miniaturized assays can be carried out with low volumes of sample in a cost-effective manner. A conventional LOC system is designed to incorporate all of the necessary steps for obtaining the information of interest on-chip. That is, the subsystems are all vertically integrated from preprocessing of the sample to detection of the signal. This perspective on design requires the adaptation and on-chip optimization of every assay that needs to be performed. The LOC approach is proving valuable for applications in diagnostics (e.g., Claros Diagnostics, Micronics, Veredus, Fluidigm), where the information of interest is tightly defined and the decisions are mostly determined based on a strict definition of a biological event (e.g., the presence of prostate-specific antigen at a certain concentration). Answering challenging questions about cell biology, however, requires a more flexible system that allows the implementation of complex bioanalytical processes that enable multidimensional, high-content analysis of single cells. In this area, LOC systems remain underdeveloped.

Comprehensive characterization of the phenotypic, functional, and genetic variations for individual cells requires multiple different measurements that can each be correlated to each cell analyzed. Given the intrinsic differences among biological traits of interest (e.g., DNA sequences and type of protein expressed on the surface of a cell), it is important to consider how to design a bioanalytical process that would comprise multiple operations carried out in series or in parallel to evaluate each trait of interest. (This approach to design incorporates, of course, multiplexed analysis within a particular trait, such as a collection of genes or expressed proteins, but aims to extend the dimensionality of the classes of data collected per process). Central to the implementation of such processes is a strategy to conduct multiple analytical operations on a given cell in an efficient and comprehensive manner. In this way, there is a horizontal integration of analysis that yields a multidimensional set of data for individual cells. We refer to this approach as "integrated single-cell analysis" (iSCA) (11).

The main operational processes comprising a typical iSCA system are classified here as cell isolation, data extraction, and data integration and analysis (Fig. 3). The isolated cells are common substrates, on which tools for extracting data are employed. Unlike conventional LOC systems, the requirement for detecting signals on-chip is relaxed, and the design of modular operations may involve the extraction of one or more classes of data for analysis on already established, commonly available off-chip technologies. One example is sequencing: It is not essential to replicate advanced sequencing technologies on-chip in most instances so long as specific genetic information of interest can be related back to the cell from which the genetic content was extracted. For the purposes of immune monitoring, the resulting information from a series of operations for individual cells should combine to yield a comprehensive, multidimensional view of the state or evolution of the cells available in a clinical sample. In the following sections, we focus on these operational tasks, provide examples of methods and technologies from the literature for each, and discuss challenges associated with advancing these technologies to realize processes for iSCA in immune monitoring.

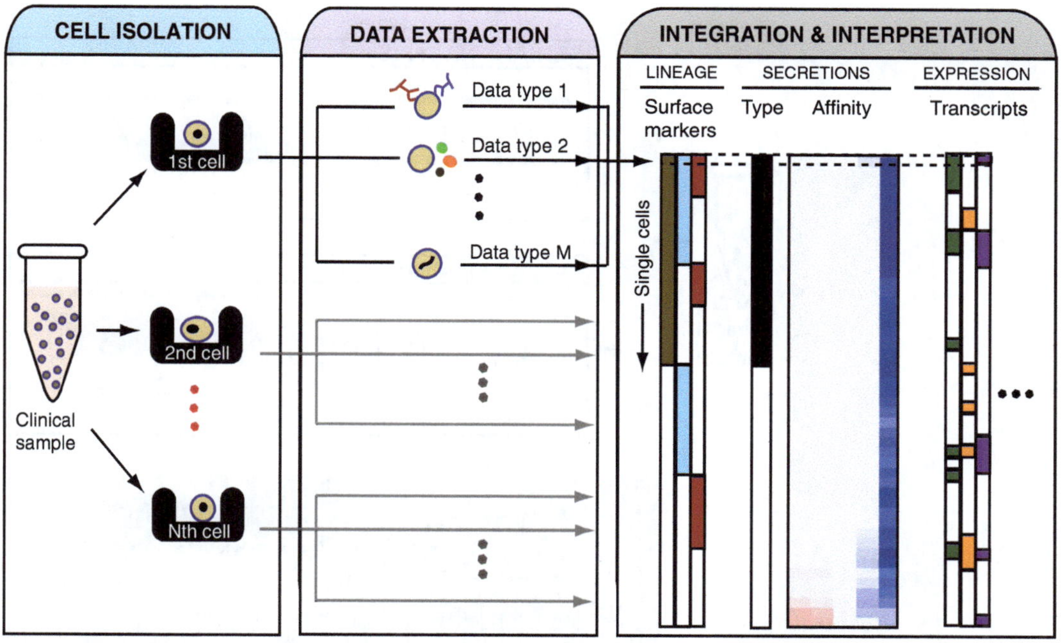

Fig. 3. Key elements of a process for integrated single-cell analysis. Individual cells are isolated from a population; alternatively, defined numbers of cells may be isolated together to allow dynamic interactions. Modular unit operations allow the extraction of multiple classes of data for each cell(s) isolated. Data for each cell are collated and integrated for analysis and interpretation to assess signatures of immune responses.

2.1. Isolating Single Cells for iSCA

The increased interest in single-cell analysis within the biology community has motivated the development of several chip-based techniques for cell isolation. A common material of choice for manufacturing devices has been poly(dimethylsiloxane) (PDMS), though other plastics and glasses are also used (12). Techniques for isolating cells can facilitate detecting signals of interest either in a serial manner (one cell at a time) or in a parallel manner (many cells at once) (Fig. 4a). In the next subsections, we provide examples of methods for isolating cells and discuss present challenges for their use in iSCA.

2.1.1. Isolating Cells for Serial Measurements

Methods for isolating cells that enable interrogation in a sequential manner include physical traps formed by actively actuated valves (13–15) and confinement within water-in-oil reverse emulsions (16–19). For devices that use active traps, cells are isolated through controlled actuation of valves and manipulation of flow in microfluidic channels. The number of connections needed per trap has limited the density of traps per device, and thus, scalability of these devices for analyzing large numbers of single cells ($>10^4$) remains an issue. For devices that use water-in-oil reverse emulsions, cells are encapsulated in “droplets” that provide a physically isolated microenvironment for each cell. These devices may limit extended

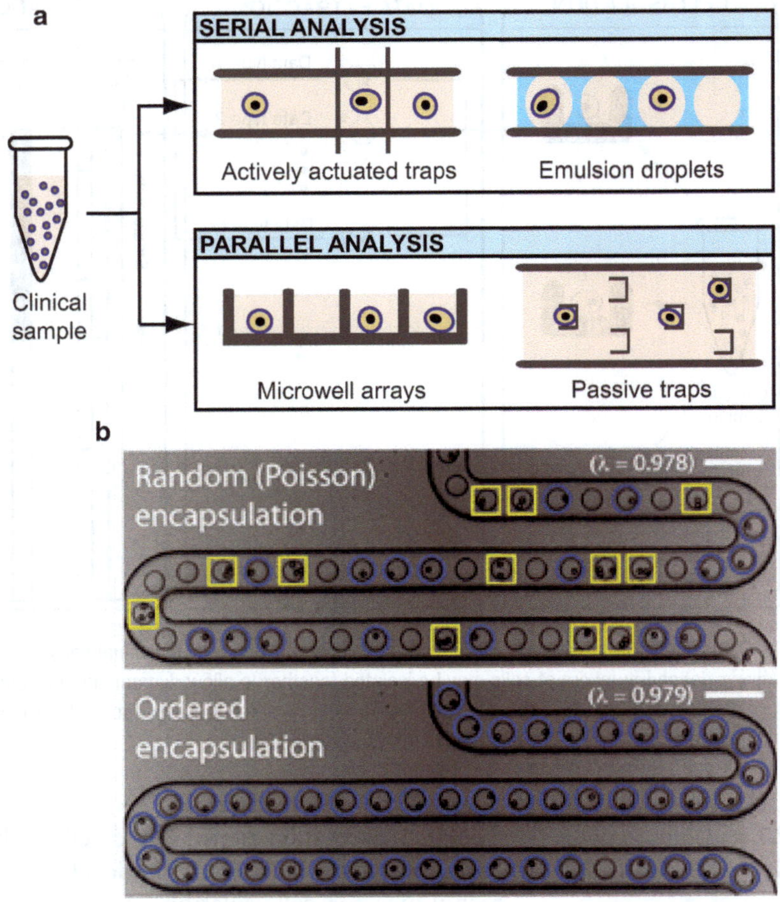

Fig. 4. Isolation of individual cells. (**a**) Examples of methods for isolating cells that permit either serial or parallel interrogation of the trapped cells. (**b**) Example micrographs of individual beads isolated in water-in-oil emulsions. (*top*) Droplets filled randomly with beads are governed by Poisson statistics, whereas (*bottom*) deterministic methods for ordering beads during formation of droplets permit higher efficiencies. Isolated single particles (*circles*), droplets with multiple particles (*squares*), and empty droplets (*unmarked*) are all observed (18). Reproduced by the permission of The Royal Society of Chemistry.

in vitro culturing (>24 h) due to the small volumes (tens of picoliters) of the droplets, and require significant manipulations to join or divide droplets to perform multistep assays. The retrieval of individual cells of interest is challenging, though integrated fluid-handling systems that incorporate feedback from fluorescence provide one means to sort droplets of interest (20, 21).

2.1.2. Isolating Cells for Parallel Measurements

Isolation of cells in a manner that facilitates simultaneous measurements on many individual cells at once is also useful. Passive cell traps (22, 23) and microstructured planar arrays (24–26) are examples that allow this approach. In passive-trap devices, cells are trapped and held in weir structures positioned in the stream of a microfluidic device, and monitored in parallel by fluorescence microscopy. One disadvantage of this method is that the fluidic environment in these devices is shared among all trapped cells. The nature of the microenvironment experienced by cells near the beginning of the channels may differ from that near the end as metabolites, secreted proteins, growth factors, and other cellular components are accumulated and washed downstream.

In a planar array of microwells, individual cells from a suspension settle down into containers with subnanoliter volumes by gravity. These devices provide a simple way of isolating cells without the need of valves, actuators, microfluidics, and external connections. Each cell is registered to a specific address within the microwell array, and therefore, may be interrogated with multiple assays, and retrieved easily by micromanipulation (27, 28).

2.1.3. Challenges for iSCA Using Existing Systems to Isolate Cells

We consider here the extent to which existing methods for isolating cells are useful for implementing iSCA. For iSCA, the method for isolation should allow the cell to be transferred among multiple analytical operations as a substrate. The ideal technique, therefore, should provide registration of each cell to allow repeated monitoring, and facilitate single-cell recovery for subsequent downstream processing.

Arrays of microwells provide the most straightforward approach for assigning individual spatially defined addresses to each cell analyzed. This address allows connections among multiple sets of extracted data for each cell to be made, and facilitates the precise recovery of cells for further analysis (29). Creating independent and controllable microenvironments in such systems would allow measuring cellular behaviors under well-defined experimental conditions of interest without any undesired perturbations from neighboring cells. Confinement of cells in droplets and active traps already provide true isolation of cells by physically separating their microenvironments from those of others.

Another important criterion, especially for identifying rare cells, is the efficiency of isolating single cells with respect to the maximum throughput of the method. This attribute determines both the sensitivity and efficiency of the method. The majority of the techniques described for isolating cells rely on loading random distributions of cells at limiting dilutions into traps. The efficiency of filling traps is governed, therefore, by Poisson statistics. To increase the number of single cells trapped, different microwell geometries have been considered (30), and the dimensions of the traps have been constrained to exclude other cells physically from the same space (25, 31, 32). A limited number of deterministic approaches have successfully improved the efficiencies for isolating single cells (Fig. 4b) (18, 33, 34), though implementing these approaches for routine use in high-throughput formats remains to be accomplished.

2.2. Methods to Extract Data from Single Cells in iSCA

To maximize the knowledge gained from a given sample, a process for iSCA should allow the extraction of many kinds of information with sufficient resolution such that multiple data points can be correlated to the same individual cells. The conventional approach for extracting data from cells in microsystems involves the co-implementation of both isolation and measurement of cells on-chip. Operations that enable the transfer of information (proteins, DNA, RNA) from isolated cells to an instrument off-chip while maintaining knowledge of its exact source provide an alternative approach for analysis. Such approaches allow the use of the most suitable and highly optimized tools available to characterize the biological element of interest, and thus, offer a significant advantage with respect to flexibility in the design of a specific analytical process. They present, however, challenges for registering generated data with individual cells.

For single-cell analysis, as reviewed in depth in this volume, important classes of biological attributes include the lineage and differentiated state of a cell (cell surface markers), secretion profile (secreted antibodies/cytokines/chemokines/growth factors), proliferation (cell size and division), cytolytic ability (cell–cell interactions), and genotype (genetic content). In the following subsections, we present examples of single-cell analytical tools that enable the extraction of these different types of information relevant for immune monitoring, and highlight which measurements are best-suited for implementing on- or off-chip (Fig. 5).

2.2.1. Differentiation of Cell Types Based on Expressed Glycoproteins

A common way to distinguish individual types of immune cells is by their surface-expressed proteins. Many glycoproteins are associated with specific subsets of immune cells, and often are referred to using a standard nomenclature (cluster of differentiation (CD)). The presence or absence of particular surface markers is used to define the lineage and differentiated state of the cell; examples include CD19 (B cells), CD3 (T cells), and CD45 (general marker of lymphocytes) (35). Flow cytometry has been and remains the most ubiquitous technology for enumerating cells based on these markers (36). In flow cytometry, individual cells are monitored within a continuous flow. As the cells pass in front of a laser, the scattered fluorescence from dye-conjugated antibodies bound to cell-surface markers is recorded. Large numbers of individual cells can be interrogated in a timely manner (20,000 cells/s) on these systems. The requirement for fluorescent labels for detection ultimately limits the multiplexing capability of this technique. Although state-of-the-art instruments allow detection of up to 17 markers, the significant overlap in spectra of the dyes requires substantial mathematical compensation to deconvolute signals (36). In flow cytometry, the detected signal from an individual cell typically is not registered to that particular cell, and therefore, with the exception of one low-throughput example (37), subsequent analysis of the same cells is not possible.

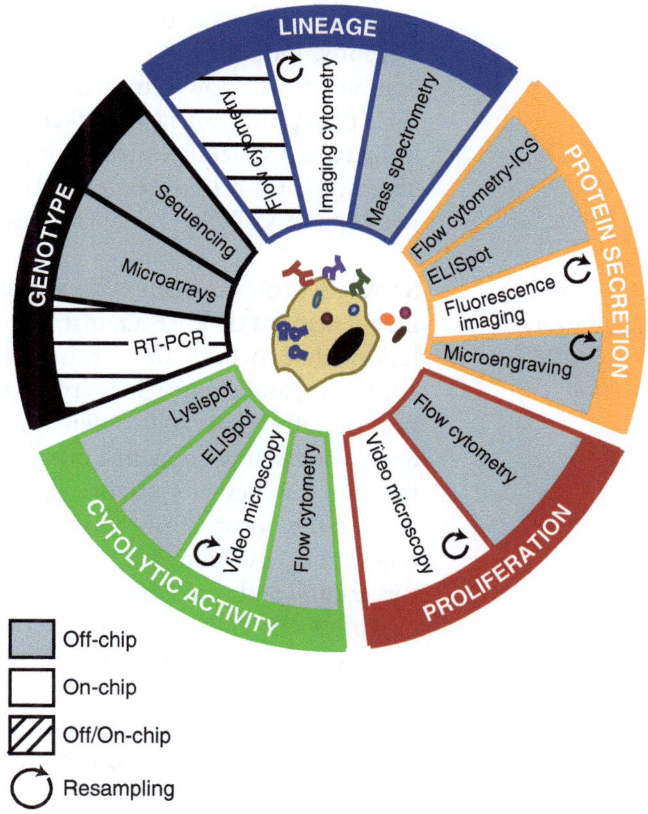

Fig. 5. Schematic overview of analytical tools commonly used to extract multiple classes of data from single cells. Tools are associated with classes of data of interest, and mode of common use indicated-off-chip (*solid-fill*), on-chip (*no-fill*), or both (patterned). Tools that allow resampling of the same cell serially are marked with a *circular arrow*.

A second approach for cytometry that is conducive to registration with multiple measurements for the same cell is image-based cytometry. This method can use either a conventional epifluorescence microscope (38) or a laser scanning cytometer (39). These instruments can take advantage of spatially patterned microsystems to assign specific addresses to individual cells. Data on the expression of certain cellular surface markers can then be linked to individual cells isolated within those wells (26, 40). Image-based cytometry has the same constraints on multiplexing that are present in flow cytometry, since both techniques require resolving the spectra of the fluorophores used. The capability for repeated use of the same cells makes fluorescence-based imaging better suited for iSCA than conventional flow cytometry, however.

A third method that has received much attention lately for characterizing proteins that differentiate cells is mass spectrometry. The use of element tags as biomarkers sets this method apart from fluorescence-based methodologies. It eliminates the issue of overlapping biomarker spectra among fluorophores, and therefore, increases the number of variables that can be monitored concurrently. Recently, inductively coupled plasma-mass spectrometry has been combined with flow cytometry, and single cells have been analyzed for a panel of 20 surface markers (41, 42). Although attractive for its increased scalability with respect to multiplexing, this method for characterizing cell-associated proteins is destructive, and as such, would necessarily be a final measurement in any process for iSCA.

2.2.2. Measurement of Secreted Proteins

Secreted proteins from cells are central to the immune system's mechanism of defense against infection, as well as to the disregulation that occurs in autoimmunity and allergy. Antibodies, cytokines, and chemokines are important for enabling the elimination of pathogens, cell-to-cell communication, and inflammation. Secretion itself is a dynamic process by nature, but determining the rate and kinetics of secretory events from single cells is not trivial. As a result, most of our existing knowledge on secretion profiles relies on data obtained by ensemble or temporally integrated measurements, and thus, only average measures of secretory behavior.

Intracellular staining (ICS) combined with flow cytometry has been widely used to quantify the numbers and relative magnitudes of cytokines produced by cells (43). In this method, chemical additives block secretion of proteins, and subsequent permeabilization of cells allows the detection of intracellular proteins. ICS has been utilized for the detection of many intracellular molecules, e.g., phosphorylation levels of intracellular signaling molecules can be determined with this method via phospho-flow cytometry (44). Blocking the secretion of proteins, however, may alter the functional response of the cell, and leaves it nonviable, prohibiting further characterization. Alternatively, technologies that use bivalent antibodies to tether secreted cytokines onto the secreting cells have been used to detect secreted proteins by flow cytometry (45).

ELISpot provides enumeration of single cells that secrete proteins of interest (6). Secreted proteins are captured on a membrane surface around the secreting cells via an ELISA-type sandwich assay. Detection of antigen-specific antibodies and cytokines has been demonstrated with this method. The technique, however, offers only a qualitative measure of the amounts of secreted proteins, and cells are lost in the process, preventing subsequent measurements (46).

Fluorescence-based imaging has also been employed to investigate cell-secreted cytokines and antibodies. Using an on-chip approach, secreted molecules bound to the immediate surrounding area of individual cells in microwell arrays have been detected. Detection of secreted interferon-γ from activated T cells (47), and screening of virus-specific antibody-secreting cells (28) has been demonstrated with this approach.

An off-chip strategy employed by our group allows the transfer of the secreted cytokines and antibodies from single cells in an array of microwells onto a glass slide without loss of registration (40, 48). This method, called microengraving, can be applied repeatedly to measure the rates of secretion for one or more proteins from the same individual cells. The resulting microarray is characterized using a conventional fluorescence-based microarray scanner. Although fluorescence-based techniques are intrinsically limited in the number of independently resolved measurements they allow due to spectral convolution, techniques such as microengraving demonstrate the utility of using off-chip detection for analysis. The method is nondestructive so increased multiplexing can be achieved through repeated interrogation to expand the breadth or dimensionality of the analysis. The capability of serially sampling the same cells allows both multiplexed analysis of antigen-specific affinities of secreted antibodies (49) and the assessment of the dynamics of protein secretion (50).

2.2.3. Assessment of Cell Proliferation

The ability to proliferate provides an additional measure of an immune cell's response to a stimulus, such as an antigen or adjuvant. The proliferation of individual immune cells is commonly characterized off-chip using a range of assays implemented with a flow cytometer (51). A classic method for identifying actively proliferating cells is incorporation of thymidine analogues, such as 5-bromo-2′-deoxyuridine (BrDU) or 5-ethynyl-2′-deoxyuridine (EdU) into newly synthesized DNA (52, 53). The incorporated analogues are stained with fluorescently labeled antibodies (for BrDU) or azides (for EdU), and the resulting fluorescence is detected on a flow cytometer to identify cells with newly synthesized DNA. One drawback of BrDU and EdU labeling is that cells must be fixed and permeabilized prior to data acquisition, and thus cannot be repeatedly observed over multiple rounds of division or recovered for downstream analysis.

Several types of stably incorporated live-cell fluorescent dyes have been developed that allow proliferation measurements to be made on viable cells. A parent population of cells is labeled with the dye, and upon each round of division, the dye is partitioned between the two daughter cells. The division history of individual cells is inferred based on the relative intensity of dye measured by flow cytometry. Cells from the parent population are brightest, whereas signals from cells of subsequent generations are diminished due to the dilution of dye through successive rounds of division. Carboxyfluorescein diacetate succinimidyl ester (CFSE) is a commonly used dye for such assays, and can track up to eight rounds of division (54). An advantage of flow cytometry-based proliferation assays is that they can be combined with simultaneous surface marker and ICS to generate integrated measurements of proliferation, phenotype, and cytokine production from single cells.

Unlike flow cytometry, video microscopy can be used to follow individual proliferating cells and their progeny over time. Video microscopy has been used to compare the time to division of both individual parent cells and their progeny, as well as to correlate morphological states such as cell size with the proliferative response of individual cells (55). Microfluidic platforms for isolating small groups of cells have also been combined with on-chip image-based cell tracking to characterize temporally regulated events, such as progression through the cell cycle (56).

There is great interest in understanding the relationship between proliferation events and other cellular attributes, such as protein expression and secretion. The combined observation of proliferation and intracellular protein expression at the single-cell level has been achieved by using live-cell microscopy to monitor cells that express fluorescent reporter proteins (4, 57). It remains a challenge, however, to correlate functional behaviors directly with a cell's history of proliferation at the single-cell level using untransfected cells and in a high-throughput manner. Our group has begun to address this challenge by developing processes for iSCA that combine an on-chip method for monitoring single-cell proliferation with an off-chip detection of protein secretion from the same individual cells (as described in Subheading 2.2.2) (50).

2.2.4. Monitoring Cytolytic Activity

Cytotoxic interactions take place when a target cell is recognized and killed by a cytolytic effector cell. Cytolytic responses mediated by effector cells, such as CD8+ T cells, natural killer (NK) cells, and macrophages play a vital role in identifying and destroying cells that are infected or malignant (58). Single-cell assays use several different approaches to quantify the killing interactions between effector and target cells. Some assays detect death of a target cell, other assays detect degranulation by effector cells, and still others make combined observations of target cells and effector cells in individual target–effector pairs.

Individual dead target cells can be identified by flow cytometry using fluorescent stains against different markers for cell death. Cells undergoing apoptosis are identified by annexin V, which labels phosphatidylserine that translocates to the outer cell membrane early during an apoptotic event (59). Cells that have undergone nuclear fragmentation can be identified with membrane-impermeant DNA intercalating dyes, such as propidium iodide (PI) (59), 7-amino-actinomycin D (7-AAD) (60), or SYTOX (61). One limitation of these markers is that they do not distinguish between target cells that died in a cytolytic interaction and those that died from other causes. This limitation can be addressed by using fluorogenic caspase (62) or Granzyme B (63) substrates to more specifically label target cells that have received a lytic hit from an effector cell.

The lytic potential of individual effector cells is commonly monitored using surrogate markers for cytolytic activity. Surface expression of the degranulation marker CD107a on effector cells can be conveniently detected using flow cytometry and has been shown to associate with increased death of target cells (64, 65). Stimulated effector cells that have not been involved in a cytolytic interaction, however, can also degranulate and express surface CD107a, limiting the specificity of the detection under certain conditions.

Flow cytometry can also be used to quantify the amount of perforin (a lytic protein) that is stored inside effector cells. Intracellular perforin expression in resting cells provides a measure of the killing *potential* of individual effector cells. However, the *realized* killing ability of individual effector cells after contact with target cells is difficult to assess based on intracellular perforin expression, since perforin is released during a cytolytic event and gradually resynthesized at rates that can vary greatly among individual cells (66).

As an alternative to flow cytometry, secretion-based ELISpot assays have been developed to directly measure the release (instead of intracellular production) of cytotoxic proteins, such as perforin (67) and Granzyme B (68), from individual effector cells. Secretion-based ELISpot assays can provide a more direct measure of *realized* cytolytic functionality, but unlike flow cytometry they are not able to concurrently measure the surface marker phenotype or multiplexed cytokine production by effector cells.

A great volume of information about the susceptibility of target cells to cytolysis imparted by effector cells can be gathered using the techniques discussed above. To completely characterize individual cytolytic interactions at the single-cell level, however, it is optimal to collect integrated measurements describing both the target and effector cells that participate in the interaction. Video microscopy is often employed to track the dynamics of interacting target and effector cells, and can generate data on interaction parameters including the number of target cells killed by each effector cell (69) and the time of contact between each target and effector cell (70). Video microscopy can also be used in combination with cell-loaded microwell arrays to monitor cytolytic interactions in many small groups of effector cells and target cells, or to monitor other individual cell–cell interactions of interest (71, 72). For flow cytometry-based measurements, one option for examining individual target–effector pairs is to focus the analyses on conjugates (60). However, this strategy cannot be used to identify target and effector cells that interacted, but then dissociated prior to the cytometry. A dual Lysispot/ELISPot secretion-based assay has also been developed to enable the concurrent detection of target cell lysis (Lysispot) and effector cell interferon-γ secretion (ELISpot) for individual target–effector interactions (73).

Integrated single-cell assays for quantifying cytolytic interactions should be capable of monitoring not only target cell death, but also other functional and phenotypic metrics, such as effector cell activation, cytokine secretion, and interaction dynamics. To accomplish these objectives, both on-chip and off-chip approaches may be required.

2.2.5. Detection of Gene Expression

Profiling expression of genes in single cells is important for understanding molecular mechanisms that regulate functional responses by cells (74, 75). Reverse transcription polymerase chain reaction (RT-PCR) has been the gold standard for the detection and quantification of cellular gene expression through conversion of mRNA into amplified cDNA. To enable single-cell RT-PCR in high-throughput formats and to better amplify low copies of mRNA, on-chip approaches that include microfluidic systems (76), microwell arrays (77), and water-in-oil reverse emulsions (78) have been employed.

Various commercial technologies are now available for high-throughput sequencing of both genomic DNA and RNA. Examples of bead-based methods include 454 pyrosequencing (Roche Diagnostics) that uses emulsion PCR for amplification of DNA attached to beads that are later read out with the help of sequencing enzymes; and SOLID sequencing (Applied Biosystems), which relies on sequencing by ligation of emulsion PCR amplified DNA captured on beads. Other examples are reversible dye-terminator-based sequencing (Illumina), which is based on imaging fluorescently labeled nucleotides as they extend surface-bound DNA base-by-base; and microarrays (Affymetrix) for sequencing by hybridization, where signal from fluorescently labeled DNA hybridized to an array with known sequences is read out. The technologies listed above have been well optimized for off-chip use, and successfully commercialized (79). They allow processing of genetic content from only small numbers of cells at a time, however, and their employment in iSCA for analyzing large numbers of discrete cells remains a challenge.

2.2.6. Challenges for Extracting Data

The ability to monitor a small number of cells of interest requires a high sensitivity for distinguishing individual cells from a large, heterogeneous population. The frequencies of antigen-specific cells of interest when investigating an immune response are typically very low (~1 in 10,000) (80). Flow cytometry offers detection of cells in 1:1,000 frequencies, and detection of lower frequencies requires special equipment and extensive procedures and acquisition times (81). On the other hand, detection of cells in frequencies as low as 1:100,000 have been demonstrated with ELISpot and fluorescent-based tools that utilize microwell arrays and droplets (48, 82, 83).

For single-cell analysis, the measured physical entities (e.g., surface markers, secreted proteins, mRNA) are often present in small quantities. Consequently, the data extraction methods need to provide the required sensitivities. Strategies that benefit from hardware technology for increasing signal collection efficiency (e.g., advanced optics, photomultiplier tubes, electron multiplying CCDs) and from signal amplification prior to detection (e.g., PCR, indirect assays with multiple fluorescent reporters) are commonly employed.

The limited viability of cells, especially primary ones, during iSCA typically sets an upper boundary condition on the time allotted for a set of modular operations (hours to days depending on the cells). Additionally, monitoring dynamic responses requires time-resolved measurements with scales defined by the molecular mechanism under study. These criteria impose a limit on the achievable sensitivities due to the tradeoff between sensitivity to dynamics and the period of time allowed for measurement. In high-throughput systems, for a given sensitivity, the maximum achievable time resolution is also constrained by the speed of the hardware components (e.g., positioning stage, camera, shutters).

One bottleneck for the currently available tools for single-cell analysis is the limitation on the number of parameters that can be scored per assay. Highly multiplexed analyses will provide a better understanding for the molecular mechanisms involved in immune responses, and allow the transition from hypothesis-driven to discovery-oriented analyses. In fluorescence-based techniques, a total of four parameters (in four channels) are commonly measured due to the choices of dyes with minimal spectral overlap. When on-chip multiplexed analysis is necessary, there is a trade-off between the numbers of channels used for each class of data (e.g., cell type and protein secretion). If the information of interest is transferred to an off-chip platform (e.g., via microengraving), the multiplexing capability (number of channels) will not be shared between different assays, and therefore provides minimally a twofold improvement in the number of parameters scored. Through repeated transfer and monitoring, the number of parameters that can be scored per assay may be increased further.

2.3. Analytical Tools for Data Integration and Interpretation

A process for iSCA should have the potential to collect multivariate datasets from large numbers of single cells. To fully utilize the rich datasets produced by iSCA, efficient strategies for data integration and analysis are needed. A key feature of iSCA is that diverse forms of data, collected both on-chip and off-chip, may be integrated for each individual cell. For example, each cell interrogated in a process for iSCA could be described by a dataset comprising parameters, such as the cell's expression level of s different surface markers, quantities of c different cytokines at time-points t_1 through t_n, proliferative behavior, and expression of g different genes at the endpoint of the assay. A variety of approaches exist for compiling and analyzing such datasets that consist of heterogeneous data (e.g., both continuous and discrete measurements). These approaches have been applied to integrate clinical data and biomolecular data (gene expression, cell phenotype and functional attributes, etc.) from cancer patients (84, 85), as well as to integrate transcriptional and phosphoproteomic data characterizing intracellular signaling networks (86). Similar analytical tools can be applied to integrate multivariate datasets collected by iSCA in the future.

Immune monitoring spans multiple scales, from molecular networks to cellular systems. Likewise, data from iSCA systems can be used to assess the correlation between different descriptors (phenotype, function, genotype) within single immune cells, as well as to address the cell-to-cell heterogeneity within a sample and the person-to-person heterogeneity within a population. Traditional data analysis approaches are useful for answering basic questions at each of these immunological scales and have been reviewed elsewhere (87). Techniques for interpreting multivariate data are particularly useful for iSCA. These techniques include partial least squares regression (PLSR) analysis for relating predictor variables and response variables, principal component analysis (PCA) for reducing data dimensionality by identifying summary variables (components) that capture most of the variation in the data, and cluster analysis for grouping observations (e.g., individual cells) based on their profile of measured parameters.

Immune responses are dynamic and involve the coordinated action of multiple cell types interacting in a network. Collection of multiparametric datasets for iSCA should inform the construction of network models of the immune response by providing information about individual cells' responses over multiple time points, as well as by quantifying the interactions that occur between individual cells. The analytical tools for generating network models are primarily based on statistical inference methods. In previous work, statistical inference tools, such as Bayesian network analysis, have been applied to construct intracellular signaling networks from phospho-flow cytometry data on T cells (44), and are amenable to adaptation for future use with other datasets from iSCA.

Ultimately, one goal for iSCA applied to immune monitoring is to infer relevant features of an individual's immunological state from a small clinical sample of primary cells. To accomplish this objective, methods for data analysis are needed to construct immunological signatures of clinical states of interest (healthy, diseased, vaccinated, etc.). Hierarchical clustering is one such method and has been used with datasets from gene expression microarrays to generate signatures of disease state for subtypes of lymphoma (88) and tuberculosis infections (89). Supervised learning algorithms are also useful for generating classifiers that can optimally predict the outcome of interest. For example, analytical approaches based on support vector machine algorithms have been used to predict the strength of an immune response to vaccination (7) and the response of single cells to drug treatments (90). Immunological profiling and clinical state prediction from small primary cell samples will become increasingly feasible as the capabilities for both acquiring and analyzing data by iSCA evolve.

3. Examples of iSCA

Our laboratory has begun to explore how processes for iSCA can provide comprehensive profiles for a given set of cells, where several biological attributes can be assigned to each cell. In one example, we have investigated how the secretion of heterologous proteins by *Pichia pastoris* evolves over time with respect to age and time to division (Fig. 6a) (50). A clonal culture of these yeast cells that were constitutively secreting a fragment of an antibody were isolated and deposited into an array of microwells. The number of cells per well and the relative age (based on bud scars) of each cell were determined by imaging cytometry. The rates of secretion for each individual cell were then determined over a period of 435 min by serial microengraving, and the number of cells in each well was then redetermined by imaging cytometry. These data collectively provided information

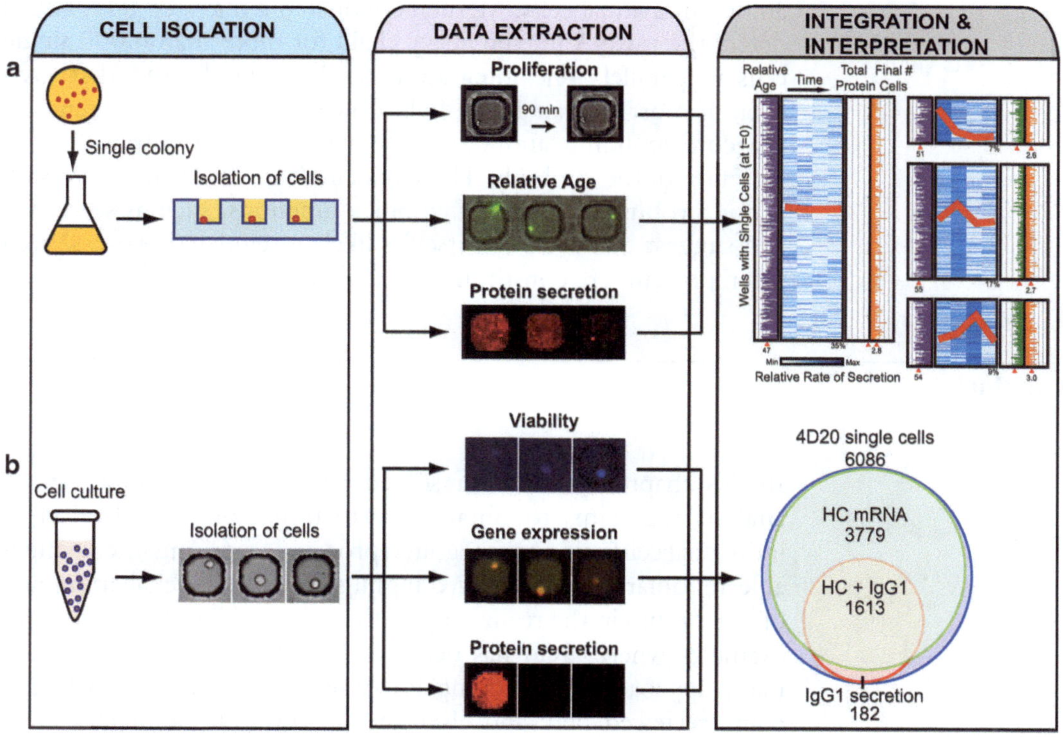

Fig. 6. Examples of demonstrated processes for iSCA. These processes include cell isolation, extraction of multidimensional data for thousands of cells using modular operations, and data integration. (**a**) A process for evaluating the secretory dynamics of heterologous protein secretion by *Pichia pastoris*. Proliferation, secretory profiles, and relative age of isolated *P. pastoris* are monitored. Adapted from ref. 50 with permission from Biotechnology and Bioengineering. (**b**) A process for assessing the relationship between gene expression and secretion of antibodies by human B cell hybridomas. Single cells were interrogated for viability, expression of their heavy chain gene, and rates of secretion. Adapted from ref. 70 with permission from Lab on a Chip.

about the average time for division for each cell, their initial relative age, and their time-dependent trajectory of secretion. We measured significant variation in the productivity among individual cells, with a substantial fraction of the cells failing to contribute (~35%). There were also dynamic fluctuations in the state of secretion ("off" and "on") that were inherited by individual cells. Interestingly, these differences did not relate to the cell's age or viability. In fact, the most productive population divided, on average, one more time than the other observed subsets. Bulk analysis of these data obscured this set of cells that appears to contribute significantly to the culture's productivity.

A second example of iSCA demonstrates the combined measurement of secreted proteins and the detection of mRNA transcripts encoding the protein for single cells (Fig. 6b) (77). Human B cell hybridomas were isolated in an array of microwells, and assessed for viability by image-based cytometry. The secretion of full-length antibodies by these cells (IgG1) was then measured by microengraving. To determine the relationship between secretion of Ig and the expression of the gene encoding the heavy chain of the antibody among individual cells, we then conducted a one-step RT-PCR to detect the mRNA for the heavy chain for more than 6,000 single cells in parallel. The integration of the data showed that even though most human B cell hybridomas transcribed the gene for the corresponding antibody, only a subset demonstrated active secretion of the antibody. These data indicate that gene expression is not a robust predictor of complex functions, such as secretion, and suggest that secretion itself is likely a regulated process, even when proteins are constitutively expressed.

4. Outlook

In this chapter, we have presented a new concept for single-cell analysis that aims to obtain information from a collection of individual cells using an integrated process comprising interchangeable modular operations. We highlighted specific challenges and opportunities in the realization of this approach, and reviewed two examples where iSCA has been implemented. Broadly, there are many approaches developing in parallel for single-cell analysis—many reviewed here and elsewhere in this volume. The specific embodiments that have the greatest impact on monitoring immune responses, and on understanding human immunology more generally, will be ones that can maximize the amount of information recovered from a given clinical sample of limited volume. Accomplishing this objective will require a shift from dedicated designs of LOC systems that detect only one or a related class of analytes to strategies for implementing fully integrated analytical processes that enable the efficient and comprehensive extraction of information from available cells. Concepts for process engineering have been applied widely in manufacturing and microelectronics, but remain underutilized for LOC technologies. These principles should guide the design and implementation of bioanalytical processes tailored to biological problems at hand, and by doing so, should provide new resolution for defining signatures of productive and successful immune responses to pathogens and therapeutic interventions alike.

References

1. Frankenstein, Z., Alon, U., and Cohen, I. R. (2006) The immune-body cytokine network defines a social architecture of cell interactions, *Biol Direct 1*, 32.
2. Fahey, J. L., Taylor, J. M. G., Detels, R., Hofmann, B., Melmed, R., Nishanian, P., and Giorgi, J. V. (1990) The Prognostic Value of Cellular and Serologic Markers in Infection with Human Immunodeficiency Virus Type-1, *New Engl J Med 322*, 166–172.
3. Shattock, R. J., Haynes, B. F., Pulendran, B., Flores, J., and Esparza, J. (2008) Improving defences at the portal of HIV entry: mucosal and innate immunity, *PLoS Med 5*, e81.
4. Spencer, S. L., Gaudet, S., Albeck, J. G., Burke, J. M., and Sorger, P. K. (2009) Non-genetic origins of cell-to-cell variability in TRAIL-induced apoptosis, *Nature 459*, 428-U144.
5. Jung, T., Schauer, U., Heusser, C., Neumann, C., and Rieger, C. (1993) Detection of intracellular cytokines by flow cytometry, *J Immunol Methods 159*, 197–207.
6. Czerkinsky, C. C., Nilsson, L. A., Nygren, H., Ouchterlony, O., and Tarkowski, A. (1983) A solid-phase enzyme-linked immunospot (ELISPOT) assay for enumeration of specific antibody-secreting cells, *J Immunol Methods 65*, 109–121.
7. Querec, T. D., Akondy, R. S., Lee, E. K., Cao, W., Nakaya, H. I., Teuwen, D., Pirani, A., Gernert, K., Deng, J., Marzolf, B., Kennedy, K., Wu, H., Bennouna, S., Oluoch, H., Miller, J., Vencio, R. Z., Mulligan, M., Aderem, A., Ahmed, R., and Pulendran, B. (2009) Systems biology approach predicts immunogenicity of the yellow fever vaccine in humans, *Nat Immunol 10*, 116–125.
8. Gaucher, D., Therrien, R., Kettaf, N., Angermann, B. R., Boucher, G., Filali-Mouhim, A., Moser, J. M., Mehta, R. S., Drake, D. R., Castro, E., Akondy, R., Rinfret, A., Yassine-Diab, B., Said, E. A., Chouikh, Y., Cameron, M. J., Clum, R., Kelvin, D., Somogyi, R., Greller, L. D., Balderas, R. S., Wilkinson, P., Pantaleo, G., Tartaglia, J., Haddad, E. K., and Sekaly, R. P. (2008) Yellow fever vaccine induces integrated multilineage and polyfunctional immune responses, *J Exp Med 205*, 3119–3131.
9. Arora, A., Simone, G., Salieb-Beugelaar, G. B., Kim, J. T., and Manz, A. (2010) Latest Developments in Micro Total Analysis Systems, *Anal Chem 82*, 4830–4847.
10. Chin, C. D., Linder, V., and Sia, S. K. (2007) Lab-on-a-chip devices for global health: Past studies and future opportunities, *Lab Chip 7*, 41–57.
11. Love, J. C. (2010) Integrated Process Design for Single-Cell Analytical Technologies, *Aiche J 56*, 2496–2502.
12. Fiorini, G. S., and Chiu, D. T. (2005) Disposable microfluidic devices: fabrication, function, and application, *Biotechniques 38*, 429–446.
13. Taylor, R. J., Falconnet, D., Niemisto, A., Ramsey, S. A., Prinz, S., Shmulevich, I., Galitski, T., and Hansen, C. L. (2009) Dynamic analysis of MAPK signaling using a high-throughput microfluidic single-cell imaging platform, *P Natl Acad Sci USA 106*, 3758–3763.
14. Thorsen, T., Maerkl, S. J., and Quake, S. R. (2002) Microfluidic large-scale integration, *Science 298*, 580–584.
15. Unger, M. A., Chou, H. P., Thorsen, T., Scherer, A., and Quake, S. R. (2000) Monolithic microfabricated valves and pumps by multilayer soft lithography, *Science 288*, 113–116.
16. Brouzes, E., Medkova, M., Savenelli, N., Marran, D., Twardowski, M., Hutchison, J. B., Rothberg, J. M., Link, D. R., Perrimon, N., and Samuels, M. L. (2009) Droplet microfluidic technology for single-cell high-throughput screening, *P Natl Acad Sci USA 106*, 14195–14200.
17. Clausell-Tormos, J., Lieber, D., Baret, J. C., El-Harrak, A., Miller, O. J., Frenz, L., Blouwolff, J., Humphry, K. J., Koster, S.,

Duan, H., Holtze, C., Weitz, D. A., Griffiths, A. D., and Merten, C. A. (2008) Droplet-based microfluidic platforms for the encapsulation and screening of mammalian cells and multicellular organisms (vol 15, pg 427, 2008), *Chem Biol 15*, 875–875.
18. Edd, J. F., Di Carlo, D., Humphry, K. J., Koster, S., Irimia, D., Weitz, D. A., and Toner, M. (2008) Controlled encapsulation of single-cells into monodisperse picolitre drops, *Lab Chip 8*, 1262–1264.
19. He, M. Y., Edgar, J. S., Jeffries, G. D. M., Lorenz, R. M., Shelby, J. P., and Chiu, D. T. (2005) Selective encapsulation of single cells and subcellular organelles into picoliter- and femtoliter-volume droplets, *Anal Chem 77*, 1539–1544.
20. Agresti, J. J., Antipov, E., Abate, A. R., Ahn, K., Rowat, A. C., Baret, J. C., Marquez, M., Klibanov, A. M., Griffiths, A. D., and Weitz, D. A. (2010) Ultrahigh-throughput screening in drop-based microfluidics for directed evolution, *Proc Natl Acad Sci USA 107*, 4004–4009.
21. Baret, J. C., Miller, O. J., Taly, V., Ryckelynck, M., El-Harrak, A., Frenz, L., Rick, C., Samuels, M. L., Hutchison, J. B., Agresti, J. J., Link, D. R., Weitz, D. A., and Griffiths, A. D. (2009) Fluorescence-activated droplet sorting (FADS): efficient microfluidic cell sorting based on enzymatic activity, *Lab Chip 9*, 1850–1858.
22. Di Carlo, D., Aghdam, N., and Lee, L. P. (2006) Single-cell enzyme concentrations, kinetics, and inhibition analysis using high-density hydrodynamic cell isolation arrays, *Anal Chem 78*, 4925–4930.
23. Skelley, A. M., Kirak, O., Suh, H., Jaenisch, R., and Voldman, J. (2009) Microfluidic control of cell pairing and fusion, *Nature Methods 6*, 147–152.
24. Love, J. C., Ronan, J. L., Grotenbreg, G. M., van der Veen, A. G., and Ploegh, H. L. (2006) A microengraving method for rapid selection of single cells producing antigen-specific antibodies, *Nat Biotechnol 24*, 703–707.
25. Rettig, J. R., and Folch, A. (2005) Large-scale single-cell trapping and imaging using microwell arrays, *Anal Chem 77*, 5628–5634.
26. Revzin, A., Sekine, K., Sin, A., Tompkins, R. G., and Toner, M. (2005) Development of a microfabricated cytometry platform for characterization and sorting of individual leukocytes, *Lab Chip 5*, 30–37.
27. Choi, J. H., Ogunniyi, A. O., Du, M. D., Du, M. N., Kretschmann, M., Eberhardt, J., and Love, J. C. (2010) Development and Optimization of a Process for Automated Recovery of Single Cells Identified by Microengraving, *Biotechnol Progr 26*, 888–895.
28. Jin, A., Ozawa, T., Tajiri, K., Obata, T., Kondo, S., Kinoshita, K., Kadowaki, S., Takahashi, K., Sugiyama, T., Kishi, H., and Muraguchi, A. (2009) A rapid and efficient single-cell manipulation method for screening antigen-specific antibody-secreting cells from human peripheral blood, *Nature Medicine 15*, 1088-U1146.
29. Di Carlo, D., and Lee, L. P. (2006) Dynamic single-cell analysis for quantitative biology, *Anal Chem 78*, 7918–7925.
30. Park, J. Y., Morgan, M., Sachs, A. N., Samorezov, J., Teller, R., Shen, Y., Pienta, K. J., and Takayama, S. (2010) Single cell trapping in larger microwells capable of supporting cell spreading and proliferation, *Microfluid Nanofluid 8*, 263–268.
31. Lee, W. C., Rigante, S., Pisano, A. P., and Kuypers, F. A. (2010) Large-scale arrays of picolitre chambers for single-cell analysis of large cell populations, *Lab Chip 10*, 2952–2958.
32. Park, M. C., Hur, J. Y., Cho, H. S., Park, S. H., and Suh, K. Y. (2011) High-throughput single-cell quantification using simple microwell-based cell docking and programmable time-course live-cell imaging, *Lab Chip 11*, 79–86.
33. Eriksson, E., Sott, K., Lundqvist, F., Sveningsson, M., Scrimgeour, J., Hanstorp, D., Goksor, M., and Graneli, A. (2010) A microfluidic device for reversible environmental changes around single cells using optical tweezers for cell selection and positioning, *Lab Chip 10*, 617–625.
34. Lu, Z., Moraes, C., Ye, G., Simmons, C. A., and Sun, Y. (2010) Single Cell Deposition and Patterning with a Robotic System, *PLoS One 5*, -.
35. Zola, H., Swart, B., Banham, A., Barry, S., Beare, A., Bensussan, A., Boumsell, L., Buckley, C. D., Buhring, H. J., Clark, G., Engel, P., Fox, D., Jin, B. Q., Macardle, P. J., Malavasi, F., Mason, D., Stockinger, H., and Yang, X. F. (2007) CD molecules 2006 - Human cell differentiation molecules, *J Immunol Methods 319*, 1–5.
36. Perfetto, S. P., Chattopadhyay, P. K., and Roederer, M. (2004) Innovation - Seventeen-colour flow cytometry: unravelling the immune system, *Nat Rev Immunol 4*, 648-U645.
37. Sitton, G., and Srienc, F. (2009) Flow Cytometry Without Alignment of Collection Optics, *Cytom Part A 75A*, 990–998.
38. Song, Q., Han, Q., Bradshaw, E. M., Kent, S. C., Raddassi, K., Nilsson, B., Nepom, G. T., Hafler, D. A., and Love, J. C. (2010) On-Chip Activation and Subsequent Detection of Individual Antigen-Specific T Cells, *Anal Chem 82*, 473–477.

39. Harnett, M. M. (2007) Laser scanning cytometry: understanding the immune system in situ, *Nat Rev Immunol 7*, 897–904.
40. Ogunniyi, A. O., Story, C. M., Papa, E., Guillen, E., and Love, J. C. (2009) Screening individual hybridomas by microengraving to discover monoclonal antibodies, *Nat Protoc 4*, 767–782.
41. Bandura, D. R., Baranov, V. I., Ornatsky, O. I., Antonov, A., Kinach, R., Lou, X. D., Pavlov, S., Vorobiev, S., Dick, J. E., and Tanner, S. D. (2009) Mass Cytometry: Technique for Real Time Single Cell Multitarget Immunoassay Based on Inductively Coupled Plasma Time-of-Flight Mass Spectrometry, *Anal Chem 81*, 6813–6822.
42. Tanner, S. D., Bandura, D. R., Ornatsky, O., Baranov, V. I., Nitz, M., and Winnik, M. A. (2008) Flow cytometer with mass spectrometer detection for massively multiplexed single-cell biomarker assay, *Pure Appl Chem 80*, 2627–2641.
43. Jung, T., Schauer, U., Heusser, C., Neumann, C., and Rieger, C. (1993) Detection of Intracellular Cytokines by Flow-Cytometry, *J Immunol Methods 159*, 197–207.
44. Sachs, K., Perez, O., Pe'er, D., Lauffenburger, D. A., and Nolan, G. P. (2005) Causal protein-signaling networks derived from multiparameter single-cell data, *Science 308*, 523–529.
45. Assenmacher, M., Lohning, M., and Radbruch, A. (2002) Detection and isolation of cytokine secreting cells using the cytometric cytokine secretion assay, *Curr Protoc Immunol Chapter 6*, Unit 6 27.
46. Streeck, H., Frahm, N., and Walker, B. D. (2009) The role of IFN-gamma Elispot assay in HIV vaccine research, *Nat Protoc 4*, 461–469.
47. Zhu, H., Stybayeva, G., Silangcruz, J., Yan, J., Ramanculov, E., Dandekar, S., George, M. D., and Revzin, A. (2009) Detecting Cytokine Release from Single T-cells, *Anal Chem 81*, 8150–8156.
48. Han, Q., Bradshaw, E. M., Nilsson, B., Hafler, D. A., and Love, J. C. (2010) Multidimensional analysis of the frequencies and rates of cytokine secretion from single cells by quantitative microengraving, *Lab Chip 10*, 1391–1400.
49. Story, C. M., Papa, E., Hu, C. C. A., Ronan, J. L., Herlihy, K., Ploegh, H. L., and Love, J. C. (2008) Profiling antibody responses by multiparametric analysis of primary B cells, *P Natl Acad Sci USA 105*, 17902–17907.
50. Love, K. R., Panagiotou, V., Jiang, B., Stadheim, T. A., and Love, J. C. (2010) Integrated single-cell analysis shows Pichia pastoris secretes protein stochastically, *Biotechnol Bioeng 106*, 319–325.
51. Nagorsen, D., Marincola, F. M., Hodgkin, P., Hawkins, E., Hasbold, J., Gett, A., Deenick, E., Todd, H., and Hommel, M. (2005) Monitoring T Cell Proliferation, In *Analyzing T Cell Responses*, pp 123–141, Springer Netherlands.
52. Rothaeusler, K., and Baumgarth, N. (2007) Assessment of cell proliferation by 5-bromodeoxyuridine (BrdU) labeling for multicolor flow cytometry, *Curr Protoc Cytom Chapter 7*, 7.31.31-37.31.31.
53. Yu, Y., Arora, A., Min, W., Roifman, C. M., and Grunebaum, E. (2009) EdU incorporation is an alternative non-radioactive assay to ((3)H) thymidine uptake for in vitro measurement of mice T-cell proliferations, *J Immunol Methods 350*, 29–35.
54. Quah, B. J., Warren, H. S., and Parish, C. R. (2007) Monitoring lymphocyte proliferation in vitro and in vivo with the intracellular fluorescent dye carboxyfluorescein diacetate succinimidyl ester, *Nat Protoc 2*, 2049–2056.
55. Hawkins, E. D., Markham, J. F., McGuinness, L. P., and Hodgkin, P. D. (2009) A single-cell pedigree analysis of alternative stochastic lymphocyte fates, *Proc Natl Acad Sci USA 106*, 13457–13462.
56. Albrecht, D. R., Underhill, G. H., Resnikoff, J., Mendelson, A., Bhatia, S. N., and Shah, J. V. (2010) Microfluidics-integrated time-lapse imaging for analysis of cellular dynamics, *Integr Biol (Camb) 2*, 278–287.
57. Rowat, A. C., Bird, J. C., Agresti, J. J., Rando, O. J., and Weitz, D. A. (2009) Tracking lineages of single cells in lines using a microfluidic device, *P Natl Acad Sci USA 106*, 18149–18154.
58. Russell, J. H., and Ley, T. J. (2002) Lymphocyte-mediated cytotoxicity, *Annu Rev Immunol 20*, 323–370.
59. Aubry, J. P., Blaecke, A., Lecoanet-Henchoz, S., Jeannin, P., Herbault, N., Caron, G., Moine, V., and Bonnefoy, J. Y. (1999) Annexin V used for measuring apoptosis in the early events of cellular cytotoxicity, *Cytometry 37*, 197–204.
60. Kim, G. G., Donnenberg, V. S., Donnenberg, A. D., mGooding, W., and Whiteside, T. L. (2007) A novel multiparametric flow cytometry-based cytotoxicity assay simultaneously immunophenotypes effector cells: comparisons to a 4 h 51Cr-release assay, *J Immunol Methods 325*, 51–66.
61. Wlodkowic, D., Skommer, J., McGuinness, D., Faley, S., Kolch, W., Darzynkiewicz, Z., and Cooper, J. M. (2009) Chip-based dynamic real-time quantification of drug-induced cytotoxicity in human tumor cells, *Anal Chem 81*, 6952–6959.

62. Liu, L., Chahroudi, A., Silvestri, G., Wernett, M. E., Kaiser, W. J., Safrit, J. T., Komoriya, A., Altman, J. D., Packard, B. Z., and Feinberg, M. B. (2002) Visualization and quantification of T cell-mediated cytotoxicity using cell-permeable fluorogenic caspase substrates, *Nat Med 8*, 185–189.
63. Packard, B. Z., Telford, W. G., Komoriya, A., and Henkart, P. A. (2007) Granzyme B activity in target cells detects attack by cytotoxic lymphocytes, *J Immunol 179*, 3812–3820.
64. Alter, G., Malenfant, J. M., and Altfeld, M. (2004) CD107a as a functional marker for the identification of natural killer cell activity, *J Immunol Methods 294*, 15–22.
65. Betts, M. R., Brenchley, J. M., Price, D. A., De Rosa, S. C., Douek, D. C., Roederer, M., and Koup, R. A. (2003) Sensitive and viable identification of antigen-specific CD8+ T cells by a flow cytometric assay for degranulation, *J Immunol Methods 281*, 65–78.
66. Hersperger, A. R., Makedonas, G., and Betts, M. R. (2008) Flow cytometric detection of perforin upregulation in human CD8 T cells, *Cytometry A 73*, 1050–1057.
67. Zuber, B., Levitsky, V., Jonsson, G., Paulie, S., Samarina, A., Grundstrom, S., Metkar, S., Norell, H., Callender, G. G., Froelich, C., and Ahlborg, N. (2005) Detection of human perforin by ELISpot and ELISA: ex vivo identification of virus-specific cells, *J Immunol Methods 302*, 13–25.
68. Shafer-Weaver, K., Sayers, T., Strobl, S., Derby, E., Ulderich, T., Baseler, M., and Malyguine, A. (2003) The Granzyme B ELISPOT assay: an alternative to the 51Cr-release assay for monitoring cell-mediated cytotoxicity, *J Transl Med 1*, 14.
69. Bhat, R., and Watzl, C. (2007) Serial killing of tumor cells by human natural killer cells–enhancement by therapeutic antibodies, *PLoS One 2*, e326.
70. Jenkins, M. R., La Gruta, N. L., Doherty, P. C., Trapani, J. A., Turner, S. J., and Waterhouse, N. J. (2009) Visualizing CTL activity for different CD8+ effector T cells supports the idea that lower TCR/epitope avidity may be advantageous for target cell killing, *Cell Death Differ 16*, 537–542.
71. Faley, S., Seale, K., Hughey, J., Schaffer, D. K., VanCompernolle, S., McKinney, B., Baudenbacher, F., Unutmaz, D., and Wikswo, J. P. (2008) Microfluidic platform for real-time signaling analysis of multiple single T cells in parallel, *Lab Chip 8*, 1700–1712.
72. Guldevall, K., Vanherberghen, B., Frisk, T., Hurtig, J., Christakou, A. E., Manneberg, O., Lindstrom, S., Andersson-Svahn, H., Wiklund, M., and Onfelt, B. (2010) Imaging immune surveillance of individual natural killer cells confined in microwell arrays, *PLoS One 5*, e15453.
73. Snyder, J. E., Bowers, W. J., Livingstone, A. M., Lee, F. E., Federoff, H. J., and Mosmann, T. R. (2003) Measuring the frequency of mouse and human cytotoxic T cells by the Lysispot assay: independent regulation of cytokine secretion and short-term killing, *Nat Med 9*, 231–235.
74. Elowitz, M. B., Levine, A. J., Siggia, E. D., and Swain, P. S. (2002) Stochastic gene expression in a single cell, *Science 297*, 1183–1186.
75. Stahlberg, A., and Bengtsson, M. (2010) Single-cell gene expression profiling using reverse transcription quantitative real-time PCR, *Methods 50*, 282–288.
76. Marcus, J. S., Anderson, W. F., and Quake, S. R. (2006) Microfluidic single-cell mRNA isolation and analysis, *Anal Chem 78*, 3084–3089.
77. Gong, Y. A., Ogunniyi, A. O., and Love, J. C. (2010) Massively parallel detection of gene expression in single cells using subnanolitre wells, *Lab Chip 10*, 2334–2337.
78. Nakano, M., Nakai, N., Kurita, H., Komatsu, J., Takashima, K., Katsura, S., and Mizuno, A. (2005) Single-molecule reverse transcription polymerase chain reaction using water-in-oil emulsion, *J Biosci Bioeng 99*, 293–295.
79. Mardis, E. R. (2008) The impact of next-generation sequencing technology on genetics, *Trends Genet 24*, 133–141.
80. Petrovsky, N., and Harrison, L. C. (1995) Cytokine-based human whole blood assay for the detection of antigen-reactive T cells, *J Immunol Methods 186*, 37–46.
81. Helms, T., Boehm, B. O., Asaad, R. J., Trezza, R. P., Lehmann, P. V., and Tary-Lehmann, M. (2000) Direct visualization of cytokine-producing recall antigen-specific CD4 memory T cells in healthy individuals and HIV patients, *J Immunol 164*, 3723–3732.
82. Guerkov, R. E., Targoni, O. S., Kreher, C. R., Boehm, B. O., Herrera, M. T., Tary-Lehmann, M., Lehmann, P. V., and Schwander, S. K. (2003) Detection of low-frequency antigen-specific IL-10-producing CD4(+) T cells via ELISPOT in PBMC: cognate vs. nonspecific production of the cytokine, *J Immunol Methods 279*, 111–121.
83. Zeng, Y., Novak, R., Shuga, J., Smith, M. T., and Mathies, R. A. (2010) High-performance single cell genetic analysis using microfluidic emulsion generator arrays, *Anal Chem 82*, 3183–3190.

84. Mlecnik, B., Sanchez-Cabo, F., Charoentong, P., Bindea, G., Pages, F., Berger, A., Galon, J., and Trajanoski, Z. (2010) Data integration and exploration for the identification of molecular mechanisms in tumor-immune cells interaction, *BMC Genomics 11 Suppl 1*, S7.
85. Gevaert, O., De Smet, F., Timmerman, D., Moreau, Y., and De Moor, B. (2006) Predicting the prognosis of breast cancer by integrating clinical and microarray data with Bayesian networks, *Bioinformatics 22*, e184–190.
86. Huang, S. S., and Fraenkel, E. (2009) Integrating proteomic, transcriptional, and interactome data reveals hidden components of signaling and regulatory networks, *Sci Signal 2*, ra40.
87. Genser, B., Cooper, P. J., Yazdanbakhsh, M., Barreto, M. L., and Rodrigues, L. C. (2007) A guide to modern statistical analysis of immunological data, *BMC Immunol 8*, 27.
88. Alizadeh, A. A., Eisen, M. B., Davis, R. E., Ma, C., Lossos, I. S., Rosenwald, A., Boldrick, J. C., Sabet, H., Tran, T., Yu, X., Powell, J. I., Yang, L., Marti, G. E., Moore, T., Hudson, J., Jr., Lu, L., Lewis, D. B., Tibshirani, R., Sherlock, G., Chan, W. C., Greiner, T. C., Weisenburger, D. D., Armitage, J. O., Warnke, R., Levy, R., Wilson, W., Grever, M. R., Byrd, J. C., Botstein, D., Brown, P. O., and Staudt, L. M. (2000) Distinct types of diffuse large B-cell lymphoma identified by gene expression profiling, *Nature 403*, 503–511.
89. Berry, M. P., Graham, C. M., McNab, F. W., Xu, Z., Bloch, S. A., Oni, T., Wilkinson, K. A., Banchereau, R., Skinner, J., Wilkinson, R. J., Quinn, C., Blankenship, D., Dhawan, R., Cush, J. J., Mejias, A., Ramilo, O., Kon, O. M., Pascual, V., Banchereau, J., Chaussabel, D., and O'Garra, A. (2010) An interferon-inducible neutrophil-driven blood transcriptional signature in human tuberculosis, *Nature 466*, 973–977.
90. Loo, L. H., Wu, L. F., and Altschuler, S. J. (2007) Image-based multivariate profiling of drug responses from single cells, *Nat Methods 4*, 445–453.

INDEX

Sara Lindström and Helene Andersson-Svahn (eds.), *Single-Cell Analysis: Methods and Protocols*, Methods in Molecular Biology, vol. 853, DOI 10.1007/978-1-61779-567-1, © Springer Science+Business Media, LLC 2012

MIX
Papier aus verantwortungsvollen Quellen
Paper from responsible sources
FSC® C105338

If you have any concerns about our products,
you can contact us on
ProductSafety@springernature.com

In case Publisher is established outside the EU,
the EU authorized representative is:
Springer Nature Customer Service Center GmbH
Europaplatz 3, 69115 Heidelberg, Germany

Printed by Libri Plureos GmbH
in Hamburg, Germany